Praise for *Sun Po*

“We in this country are too prone to deal with the issues facing us only after the problem has reached the crisis stage. . . . I believe that *Sun Power* will help to show that energy from space is a realistic proposal and that it has great commercial potential.”

Chris Kraft
Former director, Johnson Space Center

“The time is again right to bring this very important energy option to the attention of the American public.”

Joseph P. Allen
Space Shuttle astronaut

“The contents of this book should be of interest to the nontechnical reader. . . . Nansen presents the rationale for solar power satellites in an understandable form devoid of the usual technical jargon to make the subject accessible to the public.”

Dr. Peter E. Glaser
Inventor of the solar power satellite concept

“This is a timely subject for many reasons: the growing realization of the need for energy, the enhanced environmental consciousness, and the strong . . . governmental interest in finding appropriate tasks for the defense industrial infrastructure.”

Gregg E. Maryniak
Director, International Space Power Program
International Space University

“Finally, the world can raise its hopes for a renewable, nonpolluting energy source. *Sun Power* should come . . . as good news after what has been a dismal public record on power issues. As a writer who has covered power issues for two daily newspapers, I consider this an opportune time for *Sun Power* to be published.”

Patrick Moser
Independent consultant
Formerly with the *Tri-Cities Herald*

SUN POWER

The Global Solution for the Coming Energy Crisis

Ralph Nansen

SUN POWER
The Global Solution for the Coming Energy Crisis

Cover and book design: Lynn Dale

Front cover photo courtesy of NASA, taken by the Apollo 12 crew as they left earth orbit for the second lunar landing.

ISBN 0-9647021-1-8

Printed and bound in Canada

OCEAN PRESS
P.O. Box 17386
Seattle, WA 98107
(206) 706-9811

I dedicate this book to my wife,
Phyllis, for her unquestioning
support throughout our life since we
were children growing up together.

Contents

Acknowledgments

In the years that have passed as this book was being written and rewritten there have been many who contributed ideas, support, encouragement, editing, and proofreading. First and foremost is my wife Phyllis, who gave me the encouragement to start and then lived through the agony of producing the first unsuccessful version in the early 1980s. She was there again in the 1990s when it was apparent the world was in trouble, and she convinced me I must try again. She struggled through draft after draft to correct my bad grammar and redundant statements. Her ideas and guidance are a big part of this book.

It was my editor and graphics designer, Lynn Dale, who gave it polish and brought it to a quality level that I could never have achieved.

The world must thank Dr. Peter Glaser for bringing the concept of solar power satellites into existence, and William Brown for making it practical to transmit energy without wires. They have been very helpful in the support and technical information they have given me. I'm also grateful to Chris Kraft, a supportive friend and self-proclaimed solar power satellite zealot, and to the late Jack Webb, who asked Chris and I to help him produce a "science fact" movie about solar power satellites that he didn't live to complete.

I wish to thank all of my colleagues at Boeing who worked to develop solar power satellites during the intense studies of the 1970s. They were a magnificent team. There were too many to list here, but a few that represent the others are Gordon Woodcock, study manager for the DOE/NASA contracts; Dan Gregory, conceptual designer; Jack Olson, engineer, inventive genius, and artist extraordinaire; Vince Caluori, team leader and friend who tried to keep me on the straight and narrow path; Jim Jenkins, a young tiger who didn't think anything was impossible; and Dr. Joe Gauger, physicist, economist, and friend. And especially Orlando Johnson, economist, who first identified the dramatic cost benefits of solar power satellites.

Claude McIntire spent many hours trying to keep me out of trouble and finally gave up and joined in the fun. Bill Rice pushed me into TV and radio stations and helped bring the news media to believe energy from space was possible. Tom Brownell helped open doors in Washington DC.

I am very grateful to those who have helped through the years to produce this book. Arthur Orrmont, my first literary agent, worked hard to find a publisher willing to gamble on a new author. Victoria Sanders, who recognized this book as a "call to arms." Natasha Kern, who forced me to become a writer and then worked diligently to stir an unresponsive publishing industry to publish my book. Alice Price-Knight taught me what editing was all about and gave me excellent advice. My son David, confidant and engineer, who kept pushing me to keep going, and his wife Sheri, my most critical proofreader. My colleagues from around the world who have kept the concept of solar power satellites alive through the years: Fred Koomanoff, Gregg Maryniak, Brian Erb, Ray Leonard, Lucien Deschamps, Alan Ladwig, D. Patton, Frank Little, Bruce Middleton, and Nobuyuki Kaya. Special thanks also to Brian Pankhurst, a cruising friend from England, for his enthusiastic encouragement after reading an early draft; Patrick Moser, curmudgeon and writer; Wanda Gould, classmate, school teacher, and environmentalist. Mark Dale, Hugh Davis, Denton Hanford, Bill Jury, Dick Hardy, Don Jacobs, Ed Gastineau, Joe Allen, Gerry Siebert, Frank Westgate, Winnie Lee, Dr. John Hogness, Bob and Martha Cram, and many more who have given me words of encouragement when it seemed the task would never end. Thank you all.

Ralph Nansen
May 15, 1995

SUN POWER

1

Challenge for the Twenty-First Century

The sun. Worshipped by ancient people as the basis of all life, rising in glory each morning, climbing to the apex of the sky. Brilliant in its blinding white radiance—a seething mass of gases. An atomic furnace radiating energy into the great void of space—year after year for billions of years.

The earth. A blue-and-white swirl of beauty in the cold universe and yet bathed in the sun's life-giving light since its beginning. Teeming with untold billions of humans striving for a better life and yet hurtling toward destruction in that mad quest.

What new wonders can the sun hold in store for us in the next century? What will the world be like in the twenty-first century? Will America be a prosperous, dynamic nation or will our children's children look to us and ask, "What happened to our world?"

Did people living at the end of the nineteenth century, just one hundred years ago, have any idea of the changes the twentieth century would bring? Could they have imagined tens of millions of automobiles crowding the streets of our cities and driving across this great land on broad interstate highways? How could they possibly imagine millions of people traveling through the sky in sleek jets at 500 miles an hour? Or believe that man could walk on the moon? Imagine their amazement at a modern shopping mall. Would they believe a telephone call reaching halfway around the world in an instant by way of a communications satellite? What would be their reaction as they watch me writing this book at a computer the size of a book?

What made these things possible for the twentieth century? What was it that brought such prosperity to society, allowing both leisure time and money to enjoy the great American dream?

It was energy. Or, more precisely, oil. Vast quantities of cheap oil provided not only the fuel that made automobiles and airplanes practical, but also fueled the dynamic economy that made the United States the industrial giant of the century. It created jobs, built enormous private fortunes, and helped win wars. Other energy sources—natural gas, coal, hydroelectric, and nuclear—contributed to the overall picture, but it was oil that controlled the price and led the way to prosperity for the United States in the twentieth century.

Now we must ask the question for our children: "What about the future?" Can our present energy sources sustain the increased billions of inhabitants added to the planet? Will energy-hungry developed nations seize oil-producing countries to gain control of their resources, or will oil producers become international dictators? Are there enough reserves of natural resources to sustain the world of the future? What will be the price of fuel? What will be the terrible price of the ever-increasing cocoon of carbon dioxide in the atmosphere warming the earth? Will burgeoning nuclear waste create vast wastelands of uninhabitable landscapes? How great will be the scourge of hopeless, starving, emerging nations as they reach out to neighboring lands in a desperate search for survival? Can our generation continue to turn sightless eyes to environmental pollution as we blindly follow a path surely leading to the destruction of the earth?

Today the United States, with less than 5% of the world's population, consumes one quarter of the world's energy production. Three-fourths of the rest of the world lives in poverty. Many are on the verge of starvation, and the population of the world is increasing at a rate of nearly a quarter-million people *per day*.

If this increasing population used energy produced by fossil fuels at the same rate as the United States, the world would soon be overcome by the by-products of combustion—atmospheric pollutants and carbon dioxide. Even now, some scientists predict that in a few decades our planet will be devastated by changing weather patterns and possibly even flooded by the melting ice cap. And we'd be draining our finite fossil-fuel resources at an alarming rate. We'd bankrupt both the breathable air and the energy reserves of our only home.

This prospect is frightening, but you may be thinking it will not happen—that it's just more doomsaying. The really frightening aspect is that it is already happening without all of the underdeveloped nations participating. World energy consumption continues to increase as nations like China gain economic strength. The air in their cities is choked with the products of combustion. The carbon dioxide level of the earth's atmosphere is increasing at an ever-accelerating rate.

What are we doing to our planet? What will be the result of our complacency?

The vision of the twenty-first century is overcast with these threatening clouds of the overwhelming problems in our world in the closing years of the twentieth century. Future historians will judge the decisions we make as we seek to solve the immediate problems of today. Will energy hold the key for the future as it did for the past?

From the beginning of time, energy has been essential to the development of civilizations. Control of fire allowed the Bronze and Iron Ages to reach great heights. Later, coal fueled the industrial revolution in England. The pace quickened with explosive development in the twentieth century as the earth yielded untold riches of oil. The world experienced unprecedented economic growth and technological development, but there were warning signs of serious trouble.

The economy of the United States peaked in 1973 and has been in decline ever since. The resulting decay in the standard of living of its people is measured by the drop in real income of its citizens. The last year of abundant low-cost oil was 1973. Today, two decades later, the United States is finally recovering from a long recession that still has much of the rest of the world in its grip. But all is not well as industries and jobs have been lost or forced to shift to less productive service jobs. Cruel damage is being inflicted on the earth and its people by changing weather patterns caused by global warming, evidence of the effects of carbon dioxide accumulating in the atmosphere from burning fossil fuels.

What can be done to fix the long-term economy? What can be done to stop the deterioration of the world's environment? What *must* be done? Can we continue to ignore the condition of the world we will leave to our children?

Questions are easy, answers are difficult. However, in an attempt to address all the others, let me pose one more question: What do we need to do *now* to ensure a prosperous and long-lasting world for our children, their children, and the generations to follow?

We need to change how we look at energy.

We need to find an energy source that will stop the degradation of our environment and provide ample energy necessary to support the economic development of all the people of the earth as we move into the next century.

The most serious problems of the economy and the environment have been building for many years, with deep-seated causes ingrained into the pattern of our lives. Human beings have a strong resistance to change; therefore, most proposals that could actually solve the problems are rejected because they are too difficult, will take too long to achieve results, or are too costly. A politician is reluctant to pursue an idea that will not be supported by the people since his or her political life and job depend on keeping the constituency happy, at least until the next election.

So today the solutions offered by governments only address the symptoms of the disease affecting the economy and environment. After countless government studies, temporary cuts in expenditures, and billions of dollars spent on research programs, the disease is still there, eating away the vital organs of our country and our world.

To attack the disease will require a massive effort and changes that will be difficult to initiate. In order for people to accept major change, they must first be convinced that the change will bring improvement to their lives in equal measure to the anxiety the change will cause. They must also see that the change is something that is shared and not directed at selected individuals only.

Investment in the future was an essential part of the foundation built by our ancestors to assure our future. But in the modern world of sound bites, quick profits, immediate results, and instant gratification, we have forgotten many of the lessons of the past. We only need to look around us to see institutions and structures that are the result of the investment made by our predecessors. Solving the huge problems we now face will require a change from the concept of instant gratification to an investment in the future that will provide long-term, lasting benefits for us and the generations that follow.

Many of the problems, though staggering in their proportions and complexity, can be traced to a common cause—energy that is no longer cheap and at the same time is a major cause of pollution in our environment. The solution is so simple in concept it is hard to imagine why it has not been implemented. We must develop a new energy system that provides abundant, low-cost, nonpolluting energy available for all humanity. A solution simple in concept, but so difficult to achieve. Without it, progress has been stifled for two decades, and even today there is no serious long-range energy program in this country. Without affordable energy, the underdeveloped nations look into a hopeless future of poverty and starvation.

The future of mankind is dependent on abundant, low-cost energy that will not destroy our world. There is only one known source for that energy—*solar power satellites*. Yes—energy from the sun collected as it streams past the earth by giant satellites sitting in the silence of space, covered in a mantle of silky black solar cells, intercepting the life-giving rays and sending the energy to the earth. A gift of life to humanity waiting for us to have the courage to reach up and accept its abundance and promise of hope for a world drifting towards chaos.

Why is this—these huge satellites in space—a solution? We already have solar energy on the earth and it works. Why can't we just build more solar plants on earth? Wouldn't space-based solar power be prohibitively expensive? Isn't there some other solution for our energy needs?

In order to answer those questions we need to have criteria with which to evaluate the potential solutions.

The first criterion for a major new energy source is that it must be nondepletable. All of our current fossil fuel and nuclear energy power plants use the earth's resources at a prodigious rate, and these resources will be gone sometime in the not-too-distant future. The world demand for energy is becoming so great we cannot supply it with our finite stored natural resources.

The second criterion is low cost. If the cost is not low, a new source will not be developed and the energy will not be used. This does not necessarily mean it has to be low cost in the beginning if we are willing to make an investment in the future, but it must be low cost over the long term.

The third criterion is it must be environmentally clean. We can no longer continue to pollute our world without regard to the future. We must stop the damage and start to heal the earth.

The fourth criterion is it must be available to everyone. We can no longer deny energy to the emerging nations of the earth and expect to live in peace. Eventually, abundant energy must be made available to everyone on earth. This means it must be a vast source.

The fifth and last criterion is it must be in a useable form; otherwise, it will be of little help to us.

None of the energy sources in use today can satisfy these five simple but essential criteria. They all fall short in some way. Fossil fuels are being depleted and they also add to the pollution of the earth. Nuclear power uses a depletable resource and also leaves in its wake toxic nuclear waste. Hydroelectric power is generated by a wonderful renewable source, but there are very few rivers left in the world to dam and there is growing concern over the impact dams have on the fish population. Terrestrial solar power can come close, but it will always be too costly for massive, wide-spread use because of the intermittent nature of sunlight on the earth. Even as the cost of solar cells comes down, terrestrial solar power retains some inherent problems. The sun goes down at night, clouds occasionally block the sun, and the atmosphere filters out some of the energy. As a result, terrestrial solar systems must be greatly oversized and have additional energy storage systems if they are to provide continuous energy. This is not the case when we go to space to collect solar power.

The other hope held out over the years is nuclear fusion. For the past 45 years, it has been touted as the energy source of the future that is "only 20 years away." Tens of billions of dollars have been spent on research, and nuclear fusion is now farther in the future than ever before even though it is still being heavily funded.

Only solar power satellites can meet all the criteria. What are they and why can they meet the criteria that others fail?

If we were in space looking at a solar power satellite we would see a vast, flat rectangular plane of blue-black solar cells spreading over ten square kilometers of space. Its frame, a spidery web of graceful triangular trusses, is capped at one end with what appears to be a head on a short slender neck. The neck is a swivel to give the head, a circular transmitting antenna, the freedom to move. This giant monolith shimmers in the brilliant sunlight as it circles

the world 22,300 miles above the equator in geosynchronous orbit, far from the earth's shadow. The satellite's exposure to sunlight will be eclipsed for only a few short hours each year as it passes through the shadows of the spring and fall equinoxes. The energy gathered by the solar cells on the satellite is five times as much as could be collected on earth.

The magic of this immense, stark machine circling the globe is the silent and invisible beam of energy flowing from its head toward a single spot on the earth far below. An energy beam containing a billion watts of radio-frequency energy, enough to supply electricity to a city of a million people.

The beam's destination is an oval, several kilometers across, made of rows and rows of greenhouses covered with sloping glass roofs. The glass in the greenhouse roofs contains a special magic of its own. While allowing light to pass through, antenna elements in the glass capture the energy of the beam. In an instant, the beam is converted from radio frequency energy to domesticated electricity, which is plugged into existing power grids and sent to power our lives.

The receiving antenna, or rectenna, could be much simpler, but building the antenna into the roofs of greenhouses adds elegance to the design. With the greenhouses, arid land becomes productive and producing land can have its output multiplied many times. Land required for the antennas is not lost but rather utilized to feed people.

So the power plant in space, fed with the energy of the sun, delivers its power to the people of the earth.

Is it nondepletable? Its source is the sun for as long as it shines.

Is it low cost? It has the potential of providing energy at costs as low as hydroelectric dams after it is fully developed. Solar power satellites are like hydroelectric dams. Instead of damming the waters of a river they dam the sunlight that is streaming past the earth and deliver it as useful energy to the earth. There is no cost for the sunshine, just as there is no cost for the waters that flow in our rivers. The cost of the energy is dependent on the capital cost of the satellites and the cost of maintenance. Because of the benign environment of space and lack of gravity, the satellites can be very light and built to last for many decades. They will produce five times as much electricity as an earth-based solar power plant, and the wireless energy transmission will be about 65% to 70% efficient. The key to low cost will be achieving low-cost space transportation. The technology for building completely reusable launch vehicles has been demonstrated, and when the public mandate to launch solar power satellites is established, that will provide the economic justification for their development. The potential for low-cost energy is one of the satellite's major benefits.

Is it environmentally clean? That is perhaps the greatest benefit of solar power satellites. There are no pollution products associated with the energy it generates, and only the useful energy comes to the earth. It will allow our environment to heal.

Is it available to everyone? By its very nature it will be able to make energy available to all people of the earth. The satellites can be placed all the way around the world. Geosynchronous orbit is 165,000 miles around. There is room for nearly unlimited energy-generating capacity.

Finally, is it in a form that is widely usable? The energy is delivered to the earth as electricity—the most useful form of energy known to mankind.

The possibilities of solar power satellites dwarf the amazing developments of the twentieth century—*if* we have the courage to make it happen. By going to space to gather solar energy, we can have unlimited electric power that will cost less than two cents a kilowatt hour through the twenty-first century. Today the lowest cost electricity—about three cents a kilowatt hour—is in areas that have hydroelectric dams. Much of the nation pays in the order of 10 cents a kilowatt hour—and even more in some areas.

If we continue on our current course, we will experience energy costs in excess of 70 cents a kilowatt hour before the middle of the next century. The cost of doing nothing will be staggering to every individual on earth. Our atmosphere will be choked with carbon dioxide, and nuclear waste will accumulate as a ticking time bomb. Our economy and standard of living will continue to decay, and the damage to our fragile earth will surely be fatal to human life.

We can stop the destructive pollution of our atmosphere and bring dynamic economic growth for ourselves, our children, and the underdeveloped nations of the world.

All we have to do is bravely face the needs of our children in the new millennium and demand a new source of energy for the world. I offer the solar power satellite as the logical and only solution for the future of mankind.

2

Politics and Solar Power Satellites

I first heard of solar power satellites one day in 1973. I was back in Seattle after working in New Orleans as an engineering manager for Boeing on the Saturn/Apollo lunar landing program and Space Shuttle definition studies. When I walked into the office one morning my secretary greeted me with a big smile and said, "Congratulations on your new job."

My only response was a surprised, "What new job? I'm working on the space task force."

"Didn't you hear the public address announcement this morning? You've been appointed manager of the new Design-to-Cost Laboratory, effective today."

I was stunned. I was out of the space business. I stared at her without being able to say a word and headed for the office of the company's president.

He used soothing words and appealed to my loyalty to convince me to take the job. He assured me that I was the only one with both an engineering and program-cost analysis background and it was critically important to the future business of the company to understand the Design-to-Cost discipline. The Department of Defense had announced that it was initiating the concept of planning their new programs with the idea of first establishing a cost goal and then designing the hardware to meet the goal, instead of designing the hardware and then estimating what it would cost. Ollie Boileau, who was president of Boeing Aerospace Co. at the time, wanted me to set up and manage a research group that he called the Design-to-Cost Laboratory. My job was to determine how this discipline would affect our future contract bids and how the company could integrate the concept into our new programs. After the group was organized and operating I would be able to return to the space program.

I reluctantly accepted the assignment and in that frame of mind wandered through the space engineering building to say my goodbyes.

I stopped at a young designer's desk and saw over his shoulder a sketch he was making that looked a lot like an elongated, fat, Apollo capsule, and asked, "what's that?"

"Oh, that's the big onion. I'm designing it to launch solar power satellites."

"You'd better begin at the beginning. I think I must have missed something."

"Haven't you heard about Peter Glaser's idea of generating solar energy in space and then beaming it to the ground?"

"No, I guess I haven't," was all I could reply.

"Well, he thinks that we can build giant satellites covered with solar cells to generate electricity, then convert the energy to radio frequency energy for transmission to the ground. Here's what they look like," he said as he slid a colored artist's rendering from under the pile of papers on his desk. I saw an illustration of a satellite with two flat rectangular solar arrays with a disk-shaped antenna mounted between them. There was no way to judge its size, until I noticed the note clipped to the edge: "5,000 megawatt output." I gasped in amazement. Dan laughed. “Yeah, they are big. The problem is how to launch them. Gordy thought it was a nutty idea that he could prove to be impossible, but after he started running the numbers he wasn’t so sure. Now he’s convinced it will work, and I have to design a launch vehicle for it.”

“That thing doesn’t look like a launch vehicle, it looks like a reentry capsule.”

“It’s both,” was Dan’s jubilant reply. “The rocket engines are buried in the flat end for launch, then after it gets to orbit it turns around and comes back into the atmosphere reentering like an Apollo capsule. I’ve added some small rockets that fire when it nears the water to give it a soft landing. It will float like a cork and we can reuse the whole thing.”

In five minutes, leaning over a talented engineer’s shoulder, I learned about solar power satellites. My world has not been the same since.

The Crisis Begins

When fighting first broke out later that year in the Middle East between the Arabs and Israelis in what was to be called the October War, it seemed that it would have little direct effect on the US. That assumption proved to be very wrong when the Arab oil-producing nations moved to embargo oil shipments to the United States, western Europe, and Japan in retaliation for their support of Israel. The cutoff precipitated an energy crisis that shook the very foundations of the industrialized nations.

The United States was soon in a panic as gas shortages upset people’s daily lives and prices started to climb. People demanded that the government do something. Some recommended going to war to take the oil.

I knew what I wanted to do. I recognized this as the mission of the future for space. I had worked on the development of space as a commercial venture for years and I wanted to help develop a replacement energy system for oil. I knew that system could be solar power from space.

Here I was, manager of the Boeing Design-to-Cost Lab, but my heart longed to be with my friends on the space program. Whenever I had an opportunity I used my executive position to convince Boeing management how important it was to support the space programs that could help develop solar power satellites.

It was nearly two years before I was able to turn Design-to-Cost over to my deputy and return as manager of Advanced Space Programs to work on solar power satellites.

Significant progress had been made in two critical areas toward the development of solar power satellites: efficient wireless energy transmission from space and low-cost transportation of the satellite hardware to space. To help convince people that it is possible to transmit energy without wires, Bill Brown at Raytheon set up a laboratory test, using existing hardware, that achieved 54% efficiency transmitting electricity without wires. Bill Brown's invention of a rectifying receiving antenna, which converts radio-frequency energy to electricity, makes the concept of transmitting energy from space possible.

At the same time Boeing was under contract with NASA's Johnson Space Center to define future space transportation systems requirements. This study included the design of launch vehicles that could transport solar power satellite hardware to space. The company concluded that low-cost space freighters for solar power satellites could be built.

By that time, NASA believed that the idea of solar energy from space had promise, so they put together a blue-ribbon team from all their centers to audit the concept. They found that it was technically feasible, environmentally clean, and economically competitive. The audit team identified technical issues to be studied and made recommendations for future work. After I returned to the space program in 1975, Boeing was selected as one of the contractors to develop preliminary design concepts for a solar power satellite system for NASA.

NASA normally establishes competitive study teams when they are developing new programs to help ensure the highest technical accuracy. Two or more contracts with similar work statements are administered by different NASA centers, with competing contractors performing the work. The competitors are allowed to attend each other's briefings on contract results. This unique situation of briefing your competitor creates a system of checks and balances in the development of new systems.

As part of the configuration design, our team studied different ways to generate electricity on the satellites. We established subteams to define several alternative concepts. Various configurations of solar cell arrangements were investigated. Other concepts were developed and reviewed. They were evaluated by size, weight, construction difficulty, maintenance requirements, life expectancy, and cost.

Of all the possibilities we were soon able to narrow our studies to two designs. One was based on Dr. Glaser's original proposal, which was to generate electricity on the satellites in space using silicon solar cells. The second configuration used a completely different concept of electricity generation. Instead of solar cells the design used mirrors to concentrate the sunlight into a central cavity that is called a cavity absorber. The cavity absorber is a solar furnace that heats a fluid, which in turn runs turbine engines to generate electricity. We determined that this was the more efficient conversion system, and we called it Powersat.

During this time I became a public spokesman for the idea of energy from space, and Boeing's Powersat was recognized all over the country as the energy source of the future. Even though the Powersat story created a lot of interest, the engineers were now beginning to think that silicon solar cells would be a better system. Further studies showed that solar cells with no moving parts had a higher life expectancy than the Powersat's thermal engine system with many moving parts operating at high temperatures. In the trade-off between higher life expectancy and higher efficiency, we calculated that a longer life expectancy would ultimately result in lower energy costs.

Energy Policy . . . In Transition

When Jimmy Carter became President in 1977 he proposed the formation of the Department of Energy (DOE) to consolidate all of the government energy programs and policy administration within one agency. It looked like the country was finally going to establish a comprehensive energy development program to address the energy crisis we were experiencing. The Boeing team was looking forward to being part of the nation's future.

In the mid-1970s most of the government energy development effort was concentrated on nuclear power. The Atomic Energy Commission (AEC) was the government agency responsible for commercial nuclear power development. This same agency was responsible for developing atomic bombs and nuclear power plants for navy ships and submarines. All of these activities were to be integrated into Carter's new Department of Energy. Transition periods are often confusing, and this one turned out to be particularly difficult for us. The solar power satellite program, which under NASA was developing into a project of great promise with a bold solution, soon became entangled in bureaucracy and foggy vision.

Before the Department of Energy could be formed, the administration attempted to consolidate all non-nuclear energy development programs under an existing agency, the Energy Research and Development Administration (ERDA). ERDA did not want the solar power satellite program, which they considered to be a space program, and was upset with NASA for trying to help solve the energy problem. Because of the great public interest and growing Congressional support for the solar power satellites, it had to be included in the energy program, so the Office of Management and Budget (OMB) assigned total responsibility to ERDA. Meanwhile, as ERDA tried to figure out how to spell "space," all government contracts were stopped and the program essentially came to a halt.

It took them a while to get organized, and after several months' delay, the ERDA evaluation people reached the same conclusions as the NASA audit team, which was to go ahead with further development of solar power satelites. Recognizing their lack of technical expertise in the field of space, they recommended a joint program with NASA. However, before the program could begin, ERDA was absorbed into the Department of Energy.

The Atomic Energy Commission was also destined to become part of the Department of Energy and brought with it all their nuclear programs. When all the different agencies and bureaucrats were swept together into the DOE, its charter included the responsibility for the development of atomic weapons for the military, fusion and breeder reactor development, as well as the development of a civil energy plan. The AEC had a huge budget for nuclear activities and research which they brought into the new DOE. With this large infusion of AEC funds, it was only natural that the primary interest of the new department would be nuclear research and building atomic bombs. The solar power satellite program was assigned a low priority within the civilian energy segment of the department. Fred Koomanoff was assigned to run the program, and since DOE had very little interest in energy from space, they let him run it without interference, as long as it did not threaten the other programs.

Koomanoff's group established evaluation criteria for the program that concentrated on four areas: 1) technical feasibility, 2) environmental impact, 3) societal impact, and 4) cost comparison. NASA was responsible for the technical studies, and DOE retained responsibility for the others as well as for overall program management. DOE funding for three years was $15.5 million, with NASA contributing about $4 million more for space-related technology.

At last we had an opportunity to develop the energy system of the future. Solar power satellites could provide the energy to fuel the world for as long as man existed and now the government wanted us to prove it. It was the chance of a lifetime.

DOE/NASA would select two system contractors to design and develop the overall concept of the satellites and all of the necessary supporting infrastructure, including space transportation, space assembly, and the ground-based receiving antenna. There would also be supporting technology contracts and studies to consider environmental impacts and sociological impacts. All contracts were scheduled to be completed in 1980.

This was an exciting time. I was working for the largest aerospace company in the world as program manager for solar power satellites, the world's future energy system. Now all I had to do was win the contract and prove the system would work.

I was also caught in a trap. NASA is a government agency with multiple centers that had developed over the years. The largest centers in Huntsville, Alabama, and Houston, Texas, were both created to develop the Saturn/Apollo lunar landing program. The Marshall Space Flight Center in Huntsville was formed as an offshoot from the Army's Redstone Arsenal under the leadership of the legendary Wernher Von Braun. Von Braun and his team of scientists and engineers developed the V-2 rocket for Germany during World War II and at the end of the war surrendered to US troops. They were brought to this country to help develop ballistic missiles for the Army and were later transferred to NASA to develop the Saturn launch vehicles. The Johnson Space Center in Houston was formed to develop the Apollo spacecraft. Even though there was no longer a need for two major centers after the Saturn/Apollo program, government bureaucracies never shrink once they are created, so Marshall and Johnson had to compete for tasks to justify their existence.

My organization was completing a heavy-lift launch vehicle contract with the Johnson Space Center (normally the spacecraft center) and a solar power satellite contract with the Marshall Space Flight Center (normally the launch vehicle center). There would be two contracts awarded for the new Department of Energy–funded satellite studies, and each would include both the satellite and the launch vehicles. The two NASA centers were very jealous of each other, and when they offered competing contracts, each wanted our team to bid.

Boeing was caught in the middle. All of the company's major contracts for NASA programs had been with the Marshall Space Flight Center, and the Marshall people expected our loyalty. I saw two major problems with bidding on the Marshall contract. First, it was primarily a launch vehicle center and did not have a lot of expertise in space craft. Second, their overall technical and leadership capability was not as good as the Johnson Center due to the retirement of most of Von Braun's German team after the completion of the Saturn/Apollo program. The German group had run the center with iron control and had not developed effective replacements.

On the other side was the Johnson Space Center. We would have to bid in competition with their favorite contractor, Rockwell International, but the chances were much higher that Johnson and their contractor would be the big winners if the system went into full-scale development. As only government bureaucracies can, they created a no-win situation. Our dilemma was that there were two contracts offered by two different centers for the same job. We were competing for the job with other aerospace contractors. We could win the contract but choose the losing center or choose the winning center and lose the contract. If I made the wrong choice Boeing would be out of the game.

I decided to cast our lot with the Johnson Space Center in Houston. Our team bid and won. Ironically, when Boeing announced we were going to bid on the Johnson contract, Rockwell International decided not to compete with Boeing and switched centers and won the Marshall contract. We all breathed a sigh of relief and thought nothing could stop us now. We believed we would provide the earth with nondepletable energy that would stop the pollution of our atmosphere, eliminate the need for nuclear power, and reduce the cost of electricity.

Reaching for the Sun

Everything was falling into place. Our study would be based on satellites located at geosynchronous orbit, each having an output of 5,000 megawatts of electricity—the equivalent of five nuclear power plants. These huge satellites, covered with 20 square miles of solar cells, would be placed in geosynchronous orbit—22,300 miles above the equator. A satellite in geosynchronous orbit remains over one specific place on earth. At that altitude the orbital rotational speed of the satellite is exactly the same as the speed of the earth as it rotates on its axis. In twenty-four hours the satellite makes one orbit around the earth, the same time it takes the earth to make one revolution. As a result, the satellite will always be over a fixed location on the earth. All our communications satellites are placed in geosynchronous orbits in order to service one part of the world at all times.

A satellite in geosynchronous orbit spends over 99% of the time in sunlight. This is the case because the earth's axis is tilted 23 degrees from the path it follows around the sun. As a result, the satellite passes above the shadow of the earth during summer in the northern hemisphere and below the shadow in the winter. It is this tilt of the earth's axis that causes the change of seasons. As the earth progresses on its yearly trip around the sun, summer turns to autumn. While autumn leaves are falling, the days become shorter. The earth, in its flight around the sun, is starting to lean its axis away from the sun. During the autumnal equinox period, the earth's axis is no longer tilted towards the sun, but rather forward in its path around the sun. During this time, day and night are the same length. For twenty days before and after the equinox, a geosynchronous satellite passes through the earth's shadow each night.

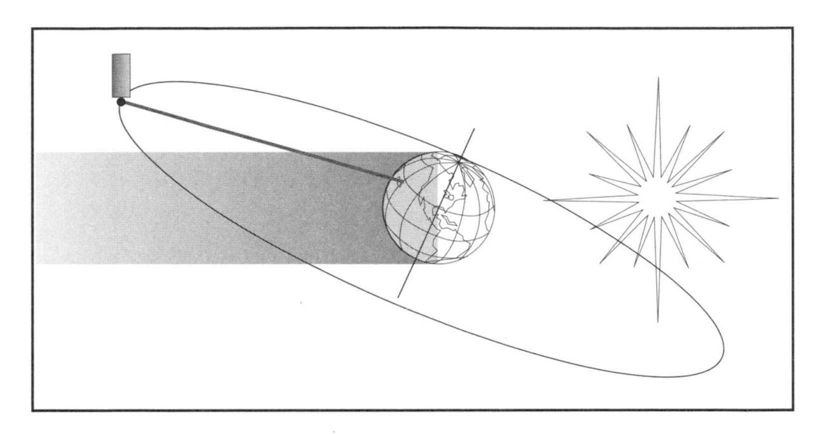

The first night the satellite will be in shadow for a minute or two. The next night it will be in shadow a couple of minutes longer, and so on until the equinox, when the maximum amount of time in shadow is 72 minutes. Then the next twenty nights follow the same schedule, but in reverse. Within twenty days of the first day of fall, the satellite will pass south of the earth's shadow and will not reenter until spring when the same phenomenon will repeat itself.

The great advantage of space solar power over land-based solar power is this continuous flow of sunlight with very little interruption. In a year it adds up to five times more energy for each solar cell in space than if that same cell was placed in the Mojave Desert—and fifteen times more than if it was placed in an average location in the United States.

The solar cells we selected for our studies were similar to the ones that have powered our communications satellites for more than three decades. They would be assembled in a giant solar array that would intercept the sunlight normally streaming past the earth.

The unique part of the concept is the transmission of power to the earth—22,300 miles away. In order to accomplish this miracle the transmitter must first convert the electrical energy gathered in space into high-frequency radio waves, which are then transmitted to the ground. The possibility of transmitting electricity without wires was actually suggested nearly a century ago by Nikola Tesla, a pioneer of the modern electrical industry. However, Guglielmo Marconi certainly did not have this in mind when he invented the first wireless telegraph in 1895, but the fundamentals are the same. Raytheon's Bill Brown finally made wireless transmission of energy work for the first time in 1964, when he flew a model helicopter powered by a radio-frequency energy beam.

Energy gathered by the satellite would be beamed to a receiver on earth. When the radio waves reach the earth they need to be captured and reconverted into electricity by a receiving antenna on the ground, called a rectenna. Electric power then flows into existing power grids in the same way as from power plants being used today. The electrical output from each satellite will be enough to power a city of millions of people.

Our task encompassed not only the preliminary design of the satellites and the energy receiver on the earth, but also space transportation systems, assembly bases, and habitats. For the hundreds of flights required to launch and build the satellites, we needed to develop a space freighter that was fully reusable and able to fly very often like today's air-cargo planes. The cost of throwing away hardware after each flight would be unacceptable.

At this point the reader might well ask, "Why couldn't you use the Space Shuttle?" And that is a good question. When Space Shuttle was first proposed, it was to be a fully reusable, winged vehicle to provide low-cost space transportation for all space programs, military and commercial. Lack of funding, congressional pork-barreling, and jealous rivalry between the NASA centers at Huntsville and Houston lead NASA to abandon their original requirements. Instead of a fully reusable, fly-back first stage, they selected the solid rocket booster, throw-away external tank concept, which resulted in only a partially reusable system. Their decision to save developmental costs resulted in a shuttle that is now prohibitively expensive to use for most of the programs for which it was designed.

I had two teams working on the space freighter design. One advocated a two-stage winged system that would fly back like an airplane and the other was a ballistic system with reentry like a space capsule. Both could provide low-cost transportation to space and were fully reusable. We finally chose the winged version over the ballistic. The big advantage was the flexibility to land on any runway and the ability to fly from place to place. They could be launched within hours of returning from their previous flight.

In the 1970s there were not many examples of robotics in use, but the engineers and scientists working on the satellites recognized that they were ideal for use in space. So we designed a system for automated assembly. All required hardware could be assembled in space by automated equipment or assembled with remotely controlled manipulators, operated by workers in the shirtsleeve environment of a control capsule. Workers in space suits would only be required in an emergency situation requiring repairs that could not be done any other way.

The contract effort was going beautifully. Boeing provided me with company funds matching the amount of contract funds. These research and development funds would help us be prepared for the follow-on development phase. I was able to spend company money to probe deeper into the areas where we could not justify spending government study money. As the studies progressed, we could see that the technology was available to do everything that needed to be done.

I had an incredible team working on the job. It was such an exciting challenge that people were beating on my office door looking for a chance to join the program. Applications poured into the company from engineers and scientists who wanted to work on solar power satellites. I could choose from the best. It was hard to turn down some of the eager young men and women who begged to be brought on board. I was pleased to hire some of the young engineers just out of school. I learned on the moon landing program that some of the best ideas came from those who had not yet learned that the impossible could not be done. I could have had an organization of ten thousand qualified engineers and scientists within a few months, if I'd had the funding.

With such a qualified group, my management job was easy. I only needed to point them in the right direction and get out of their way. This made it possible for me to become the public spokesman for the program to bring it before the American people and Congress. I received invitations to speak to various interested organizations who were anxious to find a way to solve the energy crisis. This gave me an opportunity to find out how the public felt about the idea of generating solar energy in space. The response was astounding. At the end of a presentation, audiences understood the idea and there was always a great deal of enthusiasm as they asked, "Why don't we get on with it?" and "What can we do to help?"

Congress Tries to Help

With the studies going so well, and the enthusiastic response of the public, it was only natural to think it was time to expand the development effort. However, the DOE administration was not interested in continuing the solar power satellite studies beyond the original contract. NASA could not give any follow-on contracts since their funding was controlled by DOE. The contractors appealed directly to the government for additional funding for continued development, so Washington DC was added to my travel itinerary. The first step was to brief the executive branch on the progress being made and outline how increased funding could accelerate the development. It was a discouraging effort as I found the leaders of the Department of Energy were much more interested in maintaining the status quo and protecting the nuclear industry than in supporting development of a giant competing energy source. When I talked to the leaders of the new alternative energy development organization, also in DOE, I found that their primary interest was in developing distributed energy systems, so they were also openly antagonistic to any large-scale central power concept.

Distributed systems included solar heating, wind mills, and solar cells for individual homes. To further their goal of individual energy independence, they established a single criterion for measuring the worth of any new energy system: the minimum investment cost to achieve the first kilowatt-hour of electricity delivered. This would automatically eliminate large-scale systems from serious consideration because of the high investment in infrastructure and hardware.

After getting the cold-shoulder treatment from the executive branch it was a real pleasure to brief congressmen. Usually they were much more responsive when they learned about the program. In addition to congressmen and senators, I worked with their staff members as well as the staffs of key committees. It was not long before hearings were scheduled and a bill was drafted to significantly expand the research and development effort.

During the hearings, the president of the Boeing Aerospace Company presented the formal testimony and I answered the questions. Shortly after that the solar power satellite development bill was passed by the House of Representatives. Unfortunately, the session ended before the bill came up for a vote in the Senate.

The next year we began again. We still had the support of the committee chairmen and the bill once again passed in the House. The difficulty was the Senate.

The key man was Senator Henry Jackson, the senator from Washington State, who was chairman of the Senate Energy Committee. I often talked to him about solar power satellites and he supported the concept, but told me very frankly, "Ralph, I just can't support you publicly. I am already known as the senator from Boeing and I can't add another Boeing program to the list. You have to find someone else to sponsor the bill. I'll help however I can, without going public."

A senator from Montana agreed to sponsor the bill. The real problem was getting the bill scheduled for a vote. We felt we had the votes for it to pass, if it could be brought to the floor of the

Senate. Hearings were held, and the extent of our problem became evident when the Department of Energy representative testified against its passage and indicated a Presidential veto if it was passed. Since the largest part of the budget for DOE was for breeder reactor development, fusion development, and atomic weapons development and manufacture, they effectively protected their interests by opposing any further funding for the solar power satellites.

In the meantime, public interest was building. I was now spending most of my time on the road giving speeches or briefing interested organizations. Orlando Johnson, a Boeing economist, joined the program to analyze our cost data comparisons of the future cost of electric power from the satellites to other conventional sources. The results of his analysis were startling.

During the studies, our group used cost comparisons that followed the guidelines established by NASA to compare the energy costs of various systems. It was a simplistic approach that did not take into consideration inflation costs for fuel, regulatory costs, safety and environmental escalation costs, or power plant operating time. Using these guidelines, we determined that energy from the satellites would be less expensive than any other source.

When I saw the results of Orlando's comparison, I realized that we had grossly underestimated the magnitude of the potential cost savings from energy generated by solar power satellites. He showed that with only 3% inflation and the regulatory cost escalations we had been experiencing for the last several years, the cost of energy generated by coal in the year 2040 would be 70 cents a kilowatt-hour. The cost of electricity from a solar power satellite in the same year would be less that two cents a kilowatt-hour. He went on to tell me that if we only replaced half of the United States' generating capacity by the middle of the next century the country would save *$22 trillion* with an investment of $2 trillion. The numbers were so huge they were hard to comprehend. The economic impact of solar power satellites on the future of the nation was going to be incredible.

The Ax Falls

It was now 1979. I was in a hotel room near the Los Angeles airport when I received a telephone call. I had just delivered a paper on our new energy cost comparisons to a national conference of the American Institute of Aeronautics and Astronautics (AIAA). My suitcase was packed and I was preparing to catch a flight to Washington DC after which I was scheduled to fly to Cairo, Egypt, to address the first Arab Space Conference. When I answered the phone it was Ollie Boileau, president of Boeing Aerospace, calling to tell me my trip to Egypt was canceled. I was to make no more public statements and to return immediately to Seattle. I asked why. "I had a call from Washington, and they want you to shut up," was his reply.

I went on to Washington anyway and met with one of the company's senior marketing executives and told him what was happening. His response was "you have to go, the company made a commitment to participate, let me talk to the boss, and I'll get back with you." Five hours later I was on my way to Cairo, but that was the beginning of the end.

The time finally arrived when the DOE/NASA contracts were completed and we all assembled in Lincoln, Nebraska, in April of 1980 to report on the results of the numerous studies. Represented were over 200 different organizations: the major aerospace companies and their subcontractor teams, the Environmental Protection Agency and their research scientists from universities and research institutes, concerned citizen groups representing organizations supporting the concept and groups opposing its development, research scientists from technology development companies, and economists. All had been included in the $19.5 million evaluation studies. The conclusion of the conference was that there was no technical reason why the satellite system should not be developed and that the potential benefits were very promising.

There should have been a great festive atmosphere of triumph, for the results of the studies radiated success and optimism. Instead it was like a funeral. The ax of doom hung over the proceedings. There would be no follow-on work. The contract reports were to be submitted to the Department of Energy, and at their direction, there would to be no release of the reports to the public. A new energy system was a serious threat to ongoing funding for nuclear research. The administration and the DOE wanted us out of the picture. I had been very naive to believe we could develop a new energy system that would displace coal and oil and eliminate the need for nuclear power, just because it was the best system and it would be good for the country. The opposition lined up against us was overwhelming. They were too powerful. The forces of greed had won. America and the world would suffer the consequences for years to come.

3

Our Energy Heritage

The sun is a seething mass of gases—a giant nuclear fusion reactor. An atomic furnace bathing the earth with its life-giving energy. Through the ages the earth has gathered the energy, turning some back into the void of space, converting some into the life that sets our planet apart from the others in our solar system. Some was gradually stored in the mantle of the earth's surface, some continuously stirred the fluids and gases that cover its surface. Throughout time it has been the source of all our energy.

What has it meant to us in the past? The past is important, because it can teach us the lessons that allow us to progress into the future and unravel its secrets with knowledge and understanding.

The history of mankind in the industrial age is really the history of our ability to utilize energy beyond the confines of our own bodies. Ancient people had only the strength of their arms, legs, and backs to gather food and provide shelter. Even their weapons depended on physical strength to deliver mortal blows to enemies and meat-providing animals. This meant they had to approach very closely to their prey, and occasionally it resulted in them becoming the prey if they were exceptionally ambitious about the size of their dinner. People's sphere of territory was bounded by the endurance of their legs. An individual's ability to pursue game was limited to the speed with which he could run—so he was usually forced to use skill and cunning to stalk his food, rather than speed to outdistance it. As people developed, they found it was easier to stay in one place and grow most of their food. At first, even this was very difficult because of the need to provide all the labor of cultivation with the sweat of their brow.

The Beginning—The First Energy Era

It was certainly a long time before civilization reached the point of cultivating food, but the first era of energy began long before that. It began when the first fire was lit by early man. That was the real start of civilization, for when people could control fire, they were expanding their power over energy outside of their own bodies. Fire provided warmth and protection from wild animals. It probably did not take too long until people discovered that cooking made food more palatable and provided warm food and drink on a cold night.

The fuel for the fires was wood, as it was undoubtedly the burning of forests set afire by lightning that was mankind's first experience with fire, long before people learned how to create fire by themselves. Wood was the natural fuel, as it grew nearly everywhere and was always available—it only had to be gathered together. It was the perfect fuel for the first energy era.

The Era of Wood was to last for uncounted ages, starting long before recorded history and serving human development well as civilization evolved. It was wood that provided the energy to fire pottery. It was wood-burning fires that provided the warmth that made living in the higher latitudes possible.

The invention of methods to produce fire at will must have been one of early humankind's most prized possessions. Before then, fires started by nature were carefully maintained to keep alive the precious coals so the flame could be rekindled when warmth was needed again. When the camp was moved from place to place, the glowing embers were protected with the same care as were the treasures of gold and silver in later years. If the embers were to die, the camp would have to wait for the next lightning storm to hopefully ignite a new fire. The family or tribe member who allowed the fire to die must have been dealt with severely, for the loss of fire created significant hardship—everyone would have to go without warmth, cooked food, and protection from wild animals until they could find a new source. Sustaining their fires was one of the most important elements of early people's lives.

Probably the first practical method of starting fire was by friction from rubbing dry sticks together. (Not an easy method, as I discovered when I tried to perform the task as a Boy Scout.) The individual who could start fires would have received the homage of the rest of the family or tribe. It was much later, after the start of the Iron Age, that the use of flint and iron or steel came into use. They were then used for thousands of years as civilization grew. It was not until 1827, when John Walker introduced sulfur friction, or "Lucifer" matches, that flint and steel could finally be banished as relics of the past.

Fire was the mainstay of prehistoric life, but its true importance to the development of civilization was yet to be manifested. It was the discovery of how to refine and work metals with fire that provided the advancement into first the Bronze Age and then the Iron Age. Without fire this would not have been possible. Without the ability to work metals into tools and implements, our civilization as we know it could not exist. It was about 4000 BC when copper alloys, the basis of the Bronze Age, were being used in Egypt. The Egyptians also were using fire to melt gold and silver for adornments to enrich their lives.

But it was the ability to make iron, which began about 3000 BC, that really opened the door to advanced development. With the intense heat of burning charcoal, made from wood, iron ore could be refined into iron implements, simple tools and weapons that vastly improved the quality of life for early civilizations. Ultimately, iron led to the steel that provides the basis for our advanced civilization.

As the centuries passed the use of metal spread, and by 1500 to 1000 BC the true Iron Age had begun in Syria and Palestine. By 1000 BC the use of iron had expanded to Greece, and over the next few hundred years spread throughout the known world.

Fire, with all its wonderful capabilities, did not provide any energy to aid people's mobility—that came first from the animal kingdom. The domesticating of donkeys, oxen, horses, and other beasts served transportation needs for thousands of years. Animals greatly improved the ability to cultivate fields so that one farmer could feed many families in addition to his own. Animals provided man with the ability to carry burdens far beyond the strength of his own back. They allowed people the choice of making long journeys in reasonable periods of time. However, even with this great multiplying effect, a horse could only run so fast and so far no matter how much food it was fed.

As the earth's commerce expanded to the edges of the known world, the inventive mind of human beings added a new source of energy. Wind filled the sails of ships hauling cargo and people. At first, sails supplemented oarsmen, eventually replacing them completely. As people's ability to capture and control the wind was developed, their horizons stretched to encompass the entire world. It was the energy of the wind that first made this possible.

Each step up the ladder of human development was accomplished with the aid of energy that provided the tools or materials or power that made the step possible. The use of gears led to the invention of ox-driven waterwheels for irrigation around 200 BC. The light of candles and oil lamps added to the expanding standard of living. In the year 285 AD, Pappus of Alexandria described five machines in use at that time: cogwheel, lever, pulley, screw, and wedge. These were not very complicated machines, but they were the tools that allowed people to build great buildings and to construct other machines to make life better. By the year 700 AD waterwheel-driven mills were in use throughout Europe. Fire made the manufacturing of glass possible, and glass windows came into use.

The animal kingdom and wind were great contributors to the energy pool available to our ancestors as more and more of their needs were accomplished by sources outside their physical capability, but it was still wood that was the dominant energy source during this era.

The First Energy Crisis

Through the early ages of human history, civilization developed mainly in the warmer latitudes of the earth—Syria, Greece, India, China, Egypt, Rome, and the other lands around the Mediterranean Sea. As the understanding and use of fire for warmth grew, civilization reached farther into the northern latitudes. These people were more dependent on wood to fuel their fires and provide material to build their houses, so its use was greatly increased as the population grew. The invasion of Britain by Iron Age people around 250 BC had introduced the art of iron-making to that part of the world, so additional demands were placed on the fuel supply as the use of iron for utensils, tools, and weapons expanded.

Shipbuilding became another major use of wood as England expanded its commerce across the seas to other lands. This meant a navy was required to protect the rights of the ships of commerce—and that required even more wood to build ships and feed the fires to cast the guns and forge the ships' weapons. Even the great stone cathedrals needed wood for the falsework that supported the lofty arches while they were being built. The need for wood increased so much in England that the rate of cutting down trees was higher than the rate at which they could be grown. The first energy crisis had begun.

It was a crisis that had no distinct event in history to mark its beginning. It was simply a gradual increase of demand exceeding supply. Even then it was not a universal problem, but one that evolved from one local area to another—regional shortages that at first could be supplied from neighboring areas. The lack of any large-scale transportation system would have made this expensive, however, so the first indication of the crisis must have been the lack of ready availability and the resulting increase in cost.

The period in which this first occurred is not clear, but it probably started around 1500 or possibly earlier in some areas. Its duration was to last for over a century in England. For some of the very poor nations on earth it is still going on today as they have stripped their lands of trees in the attempt to stay warm and cook their food. However, it is what happened in England that has had a lasting impact on all of the world's people.

In England, the demand for wood for heating, cooking, ironworks, and shipbuilding became so high it could no longer be supplied from within the boundaries of the British Isles. The increasing scarcity of wood meant that the price increased, so conservation was forced on those who could not afford the increased cost. Alternative fuels were used where they could be found. Alternative materials were used where practical for buildings. Wood was imported from the colonies and the European continent to supplement the diminishing supply. As a result of this added transportation cost and increased demand, the price of wood rose sharply. There was also no substitute for wood to build ships. England depended on her navy and merchant fleet for survival, so wood had to be saved for ships and other important uses—such as trees for the king's hunting forest.

The search for alternative sources of fuel was certainly not a well-organized plan of evaluation by panels of experts working under government direction. And it certainly didn't happen overnight. Rather, it is more likely that some wandering hunter found himself far from home one night and stopped to build a fire to cook his meal. He was probably shocked and wondering what kind of spell had been cast upon the black rocks—coal—that caught fire where he built his evening fire, for in England at that time coal could be found laying on the ground in some locations. It was probably used by some people for many years before it was first mined in Newcastle in 1233.

Although we do not have a date for the beginning of the first energy crisis, the year 1233 marked the beginning of the end of the era of wood. That era was to last for another 400 years, but in 1233 the stage was being set for one of the most dramatic periods in the development of civilization, as coal was to become the energy source that would fuel the industrial revolution.

As wood became scarce a combination of conditions resulted in an economic crisis with many of the same problems experienced by the modern world following the 1973-74 oil embargo. Although there was still a good supply of wood in the world, the cost of importing it damaged the British economy and living standard. Wood had been the main fuel throughout history up to this time, and even though coal was a known commodity it was not easy to make the conversion from wood to coal. Coal did not exist everywhere—it had to be transported from the mines to the users. It was difficult and hazardous to mine. It was dirty and heavy, and the smiths did not know how to make iron with coal.

By 1580 the capacity of wood to meet the needs of the expanding nation was exceeded. New buildings were banned in London to restrict growth of the city. The first energy crisis was at hand and an era—the Era of Wood, which had started when humans harnessed the first fire—was at an end.

The Second Era and the Industrial Revolution

With wood so scarce, England turned to the earth for fuel. Out of the ground came the black rocks to fire the second era of energy—the Era of Coal. Coal heated homes and cooked food, but at first the energy to run machines still came from wind or animals or people. Iron was still manufactured with charcoal made from wood.

It wasn't until 1640 that coke was first distilled from coal. From then on the iron and steel industry blossomed as new processes were developed to use the enhanced thermal properties of coal. Benjamin Huntsman improved the "crucible" process for smelting steel in 1740. The first iron-rolling mill was established at Fareham in Hampshire, England, in 1754. At the Carron ironworks in Stirlingshire, Scotland, cast iron was converted into malleable iron for the first time in 1762. Coal was providing the iron and steel for the machines needed for the industrial revolution.

Other industries in England were flourishing as well. Linen goods had been manufactured for centuries, and cotton goods were added in 1641. Sawmills were being run by wind, and paper mills had been in operation since 1590. The great void in industrialization, however, was the energy to run the machines. Horses and oxen and people and wind were not enough. The white hot heat of a coal fire was not enough by itself. What was needed was an engine to put the heat of a coal fire to work, to run the machines of industry.

James Watt is generally given the credit for inventing the steam engine, which finally made it possible for coal to achieve its full destiny, but many others made contributions along the way. Men had dreamed for centuries of using steam to produce useful work. More than a hundred years before Christ, Hero of Alexandria described a method of producing rotary motion using steam. In 1543 the Spanish navigator and mechanician Blasco da Garay submitted the design for a steam boat to King Charles V. Nearly a century later, in 1630, a French writer described a method of raising water to the upper part of a house by means of steam. A book called *Century of Inventions*, published in 1655 by the Marquis of Worcester, included a similar method. In 1690, the French engineer Denis Papin devised a pump with a piston that was raised by steam; in 1707 he invented the high-pressure boiler. Captain Savery was granted a patent in 1698 for a primitive method of utilizing the power of steam.

But it was the work of Newcomen and Cawley, who in 1705 constructed a machine with a detached steam boiler for pumping, that was ultimately used by James Watt as the foundation for his development of successful steam engines.

Watt was born in Greenlock, Scotland, in 1736 and spent his early years being taught by his mother and father at home because of poor health. Even as a young boy he was fascinated by the characteristics of steam and would sit by the hour watching the tea kettle on his mother's stove playing with the lid in the rising steam. Years later, when he was at the University of Glascow working as a mathematical instrument maker, he came across one of the Newcomen engines in the laboratory. He was soon assigned to repair it. In the process he improved its construction and made it work successfully so it could be used for instruction in the classroom. It was a very incomplete machine and not capable of doing any serious work, but it was enough for James Watt. He used it as a model to develop steam engines that could perform real work. The key was his invention in 1764 of the steam condenser, and in 1769 Watt secured his patent on the steam engine. He made steam engines that would do the work of the world—when fired with the energy of burning coal, his engines carried civilization into the industrial age. In spite of Watt's sickly youth he lived to be 83, and many of his engines lasted much longer; some are even in running condition today.

The coal mines of England were one of the first beneficiaries of coal-fired steam power. They were often subject to flooding and had to be pumped—a very difficult task until the invention of the steam engine. Many of Watt's early operational engines were designed and built to pump water from coal mines.

The development of the steam engine was the beginning of another giant step forward in man's ability to multiply his own strengths and productivity. He now had a way to convert thermal energy into mechanical energy.

Improvements came rapidly, and in 1782 Watt added his invention of the double-acting rotary steam engine. Only 25 years after Watt perfected his engine, Robert Fulton used one to power the *Clermont* from Albany to New York in less than 36 hours. No longer would man be bound by the fetters of his own strength, or that of the horse, or the whims of the wind. Man was now the master.

The pace of civilization exploded into the industrial revolution as the energy of coal-burning steam engines performed work that had previously been impossible. Industrial development blossomed. James Watt and Matthew Boulton installed the first steam engine to run a cotton-spinning factory at Papplewick, Nottinghamshire, in 1785. Manchester had its own factory by 1787. The first steam-powered rolling mills were built in England in 1790, and by 1797 England was exporting iron.

The age of rail transportation began in 1801 with the installation of the first iron trolley tracks at Croydon-Wandsworth, England. George Stephenson constructed the first practical steam locomotive at Killingsworth Collier near Newcastle in 1814; in 1822 he built the first iron railroad bridge for the Stockton-Darlington line, which opened in 1825 as the world's first railroad line to carry passengers. The Liverpool-Manchester line opened in 1830.

In the US construction was started on the Baltimore and Ohio railroad in 1828. The first French railroad line, from St. Etienne to Andrezieux, opened in 1832. The first German line opened in 1835. By that time over a thousand miles of track was in use in America. These examples were only the beginning as creative minds dreamed of new ways to put energy to work providing mobility and dramatically increased productivity. One person could control a powerful machine doing the work of large numbers of people. And because of improved transportation, ideas could now be exchanged more rapidly at greater distances.

The second era, the Era of Coal, was in full swing. It had started in England because of necessity, but it soon took on a life of its own. With its ability to power thermal engines and produce iron and steel, coal elevated England, a small island nation, to the dominant industrial power on earth. England was not the only country with coal, however, and it did not take long for the Era of Coal to spread to most of the industrialized world. Without it, other nations could not compete in the world markets. Nations that had their own coal resources were the most fortunate. As their economies expanded, their people had the highest standards of living.

The nineteenth century saw the zenith of the Era of Coal as industrial development spread throughout the world. The ships of commerce lost their sails, and their masts became smokestacks. Railroads spanned the American continent and crisscrossed Europe. Coal was not the only fuel, but it was the standard-bearer for industrial use. Wood was still used in many homes, but it could not satisfy the demands of industry.

Coal was not the perfect answer, however, as the engines it powered were not adaptable to all uses and people yearned for even more than coal could provide.

There were other problems as well. Burning coal produced prodigious amounts of pollutants, and the machinery required for mobile systems was large. As early as 1661 the effects of air pollution were attacked by John Evelyn in his writing, *Fumifugium, or The Inconvenience of the Air and Smoke of London Dissipated.* At that time, the world population was only 500 million; all of England had less than seven million inhabitants.

Unfortunately, the air did not get any better through the ensuing three centuries as the population grew. Not until after the Second World War, when London smog was so bad it killed many people, did the government finally ban the burning of coal in private homes. But in the heyday of coal, the issue of air pollution did not have a serious impact on its use because of the relatively small population and the tremendous financial benefits it was providing. The smoke-belching factory chimneys and black trails of soot streaming behind the passing trains became the symbols of wealth.

However, coal could not satisfy the need for lights, so other methods and fuels were used. As early as 1702, many German towns were lit by oil. The first attempts to use gas lighting were made in 1786 in both England and Germany. By 1807 London streets were lit by gas. Boston had gas street lights by 1822, and the Unter den Linden in Berlin had gas lights in 1826.

Other developments were taking place as well in the rich environment of industrial development. The mysterious world of electricity was first glimpsed by Stephen Gray in 1729 when he determined that some bodies conduct electricity and some are insulators; by Von Kleist in 1745 when he invented the capacitor; and then by Benjamin Franklin in his famous experiments with lightning in 1752. These early efforts were but the precursors of what was to be discovered about electricity during the nineteenth century.

Our language is filled with words like ampere, volt, ohms, faraday, henry, wheatstone, and gauss. Today they are names of electrical characteristics, but in the last century they were the names of men—scientists, engineers, and inventors who searched for and discovered the mysteries of electricity. By the start of the twentieth century the world had coal-powered and water-powered electric-generating plants, city streets were being lit with electricity, and practical batteries had been developed.

Other fuels were also tried and other kinds of engines were investigated. In 1860 the French engineer Jean-Joseph-Etienne Lenoir constructed the first practical internal combustion engine using illuminating gas as a fuel: the natural gas fuel that was used for gas lights. This was followed in 1876 by the first successful engine operating on the four-stroke principle, built by the German technician Nikolaus August Otto, also using illuminating gas as fuel. But it was another German, Karl Benz, who really inaugurated the automobile age when he obtained a patent on a self-propelled vehicle powered by a gasoline engine in 1886. He built his first four-wheel car in 1893. That same year Henry Ford built his first automobile, and the future of the world was changed forever.

At the end of the nineteenth century, the Era of Coal was at its peak. There were no shortages, but the industrial revolution it had created was simply outgrowing coal's capability to satisfy all of the expanding demands. A new, cheaper, and more flexible source was required if modern civilization was to continue to develop.

Like a tough old fighter who had struggled and risen to be world champion, then was suddenly beaten and displaced by a young kid, so king coal would be replaced by oil. The Era of Coal was heading for a sudden end, though it is still a major source of electric power generation today.

The Third Era—The Era of Oil

Even though the knowledge of oil reaches back before the start of recorded history, it did not become significant until modern times. Thousands of years ago, oil was used in various parts of the world to light lamps. It was also used as a glue under inlaid mosaic walls and floors, and Noah used it to seal his ark by applying two coats of bitumen outside and one coat inside. The Romans used flaming oil containers to destroy the Saracen fleet in 670. In the same century, the Japanese were digging wells to depths approaching 900 feet with picks and shovels in search of oil, and by 1100, the Chinese had reached depths of more than 3,000 feet, also with picks and shovels. This all happened centuries before the West had sunk its first well.

Most of the early oil was found in the form of asphalt in natural beds on the surface where oil had bubbled up out of the ground. Lighter, more volatile components had been refined away by nature, leaving the heavier fuels and asphalt. As the centuries passed, more and more oil was found naturally. Throughout the part of the world we now call the Middle East there were a great many seepages. Some were ignited by chance and became the "eternal fires" of the Persian fire worshipers. The ancient Persians used petroleum in warfare by shooting arrows tipped with burning pitch into the ranks of their enemies.

Marco Polo visited the oil fields of Baku near the end of the thirteenth century and told of "a fountain from which oil springs in great abundance, inasmuch as a hundred shiploads might be taken from it at one time," and added that "this oil is not good to use with food, but it is good to burn." Baku is on the Caspian Sea, now a part of Russia's great oil-producing fields.

During the middle of the eighteenth century, a Swede named Peter Kalm published a report of his travels in the British colonies in America. One of his finds was a place he called "Oil Springs" in an area that was to become Pennsylvania. The course of humanity was to later be changed by those springs.

By the early part of the nineteenth century, oil from Pennsylvania was selling for two dollars a gallon. By mid-century, it was down to seventy-five cents a gallon and was replacing sperm whale oil. Oil's use for lighting was spurred by the invention of kerosene, an oil refined from coal, by Dr. Abraham Gesner of Nova Scotia in 1846. It was so successful that other companies soon began making "coal-oil." This brought about a revived interest in petroleum, and in 1855 at Oil Creek, Pennsylvania, Professor Benjamin Silliman ofYale showed by experiments that petroleum was as good as oil made from coal.

Because of the growing demand for petroleum, a group of business men led by Edwin L. Drake decided to drill for oil using salt-well-drilling equipment. Their first commercial well was sunk by William A. Smith near Titusville, Pennsylvania, in 1859, and produced ten barrels a day. This more than doubled the maximum production in America at the time, which had been 2,000 barrels a year—the same as Russia, the other big oil producer.

In the latter part of the century, Pennsylvania's oil production soared and prices dropped. John D. Rockefeller founded Standard Oil Company in 1870 to take part in the growing oil industry. Stimulated by availability, low price, and its wonderful characteristics, the uses for oil mushroomed.

The golden age of oil had not yet started, however. That day was reserved for the twentieth century. It came upon us suddenly in an unexpected place. On January 10, 1901, at 10:30 a.m. in a place just south of Beaumont, Texas, a drilling crew was replacing a pipe string to continue exploratory drilling in solid rock at 1,160 feet. Suddenly, mud started flowing out of the hole and history came behind it. The gusher named "Spindletop" reached skyward for its place in history, pouring forth 100,000 barrels a day of liquid black wealth, providing people with more energy than they had ever dreamed possible. The third era of energy, the Era of Oil, had begun.

Oil made many things possible. One of these was individual transportation. Oil provided the low-cost, easily handled fuels necessary for practical internal combustion engines. This in turn led to practical automobiles and then to airplanes. Due to its ability to travel long distances, the automobile rapidly moved from the status of a rich man's toy to a necessity, becoming the principle mode of everyday transportation. A massive new industry was created to build, distribute, market, and maintain the automobile.

Oil took on the magic of "black gold," and everyone who found it acquired the Midas touch. The stink of oil refineries became the smell of money. Vast fortunes were pumped from the ground to build yet another giant industry—the business of selling energy in liquid form. It was to become the biggest business of all.

The United States economy flourished under this diet of cheap energy. Farm machines were built that greatly multiplied a farmer's productivity. Automobiles provided recreation as well as commerce. People could afford to travel long distances and see more of our vast land. Oil and its partner, natural gas, provided comfortable, clean heat for homes and industry. Process heating for industry provided a new and better approach for fabricating products. Productivity was raised higher and higher, resulting in more leisure time for the average worker. Prices were reduced. The standard of living rose continuously. It was no longer a question of scratching out a bare existence, but rather a question of how best to utilize one's free time.

The golden age of oil lasted for three-quarters of a century before its luster began to tarnish. During that time, the United States enjoyed the position of dominating the world economy. Travel was cheap, and most garages contained two cars. Americans drove instead of walked. The ability to choose whom to ride with, and where and when to travel, gave us unprecedented personal freedom. Cars were big, comfortable, and fast. Industry had total freedom to develop without serious worry about energy sources or consumption. Our homes were heated and air-conditioned to comfortable levels. As air-conditioning gave us the ability to be comfortable even in the hot, humid climates, the southern part of our nation blossomed. People had control over their environment, which gave them a choice of where to live regardless of climate. Energy costs were incidental to the general cost of living.

Even today, oil and natural gas supply 60.6% of the world's energy consumption and 65.5% of the United States' energy. It is no longer cheap, however.

The Water Wheel Comes of Age

During this golden age of low-cost energy based on oil and gas, other alternative energy systems had to be very good to compete. As is true in all fields, there is no single solution to meet all needs. This is certainly true with energy. Oil and gas provided the major energy base and established the price standard, but the emergence of electric power as a better form of energy for certain applications, such as lighting, led to the need to develop better electricity-generating methods. Key among these in some parts of the country was development of hydroelectric power plants. Two major areas were the Tennessee Valley Authority and the Bonneville Power Administration in the Pacific Northwest.

Grand Coulee Dam in Washington State was the product of a few men's dreams to convert the arid desert areas of central and eastern Washington into a garden by damming the mighty Columbia River and pumping water into a vast irrigation system. Electricity was to be a by-product. The electric power not used for pumping water was going to be sold to help defray construction costs. This dream, which my father and his partner participated in, was not easily sold to a hard-nosed Congress that could see neither the need for more food production nor how all that electricity could possibly be used. It took nearly two decades of persuasion and a deep economic depression to bring the nation to the point of committing the massive funds required. The real incentive at the time was jobs. The construction of Grand Coulee Dam drew workers from all over the nation during the Great Depression—a period when a job was one of the most precious commodities of all.

Grand Coulee and other dams in the Pacific Northwest provided an asset beyond the wildest dreams of the original visionaries. Not only was the desert turned into a garden, but electric power was produced in such quantities and at such low costs that new industries bloomed, particularly energy-intensive industries such as aluminum production. American heavy bombers pounded Germany and Japan during World War II flying on wings made from aluminum produced with the aid of low-cost electricity from hydroelectric power plants. The initial cost of the dams and generators was high, but the power produced was massive and the source, water, was free.

The story of the Tennessee Valley Authority has many similarities. With the build-up of electric power generating capacity, a quiet rural area was converted into a dynamic, growing, vital part of the nation. The availability of abundant low-cost electricity sparked economic expansion of industry and jobs and raised the basic standard of living.

Grand Coulee Dam has changed from a public works project conceived to irrigate land into one of the largest and most economical power plants in the world. However, there is limited water flow in the nation's rivers, and with 11 dams, the Columbia has given its all. The rest of the world's waterways have also been extensively developed, but the total global energy contribution of hydroelectric power is only 6.2%.

The Atomic Age Dawns

The atomic age introduced another energy competitor to the scene. Nuclear power plants became the promise of the future. The plants tapped the energy of the atom with its attendant large energy release from a very small amount of fuel. Unfortunately, the conversion from atomic heat energy to electricity involves a thermodynamic process of converting thermal energy to mechanical energy, and finally to electricity. This laborious cycle has proven to be of low efficiency and requires very sophisticated machinery. An added problem is nuclear energy's attendant radiation dangers. These dangers have taken many forms—the basic fuel has some radioactivity, and it must be processed into fuel-grade materials. It causes contamination of the machinery and working fluids in the reactor. It also poses a problem in disposing of the leftover wastes. Even so, electric power generated from a well-designed and operated atomic power plant proved to be reasonably economical. Atomic power plants now provide 5.8% of world energy use.

Coal has continued to be an energy contributor where it is economical to mine, providing lower cost electric power than other sources. Its use as a fuel has become limited primarily to electric power plants and to large industrial users who could control the environmental impact. It still provides 27.2% of the world's energy.

Energy's Gift

When we consider a country's standard of living and look back through history to search for common patterns, we soon find that there are two factors so similar that it is hard to conclude anything except that they are interrelated. These two factors are the growth of the gross national product and the growth of that country's energy consumption. The gross national product is a basic measurement of the economic viability of a nation and an indicator of the people's standard of living. Interestingly, energy growth has to occur at a higher rate than the gross national product or a nation does not progress. There are variations in the pattern caused by world events, such as the 1973 oil embargo and the heavy concentration of military hardware manufactured in the Soviet block nations in the past, but these variations tend to confirm the relationship.

With the advent of the oil embargo of 1973-74 our energy use was dramatically reduced by serious conservation measures. At first the effort was to reduce consumption by simply restricting usage. As time went by this was replaced by improved efficiencies so that many of our previous levels of endeavor could be resumed at a lower energy consumption rate. Improvements in efficiency can only be carried so far, however, until a point of diminishing returns is reached. Conservation is, after all, the organization of scarcity. Many areas of energy use have reached that point, and usage has started to grow again. When conservation is carried too far or the price continues to rise, the economy suffers. A good example today is the airline industry, which has been suffering crippling losses due largely to spiraling fuel costs. Most airlines already operate the new and very fuel-efficient planes.

The former Soviet Union offers a different case. They had a fairly high energy consumption level that did not result in a correspondingly high standard of living, because so much of the energy was oriented to building and operating a very large military force. In addition, their fuel-consuming systems were relatively inefficient in addition to being in a very cold climate.

The United States has had the highest consumption of energy per person of any of the major nations. It also has had one of the highest standards of living. Many people might argue with that statement, but in traveling the world, I have seen that although there are some nations that are comparable, no large nation exceeds the US.

However, the issue is not whether the United States has the highest standard of living. The real issue is what energy means to the rest of the countries of the world. The answer is simple. If other nations cannot have low-cost energy, they have no hope of emerging into a high standard of living and the majority of their people will remain at the subsistence level. If this situation is to change, the world's energy use has to rise to levels many times the current consumption. The United States alone has 4-1/2% of the world's population, yet consumes over *25%* of the world's energy output. If the rest of the world were to rise to only *half* the level of the United States, world energy consumption would be about *three times* the current usage. What would oil cost per barrel if it had to supply a demand of that magnitude? What hope is there for the underdeveloped nations?

4

The Great Energy Crisis

In 1973 the United States was at the peak of its economic development. It was the highest creditor nation, it had the highest per-capita gross domestic product, and real income for the average worker was at its maximum. Then Saudi Arabia led the Arab countries and other OPEC nations in the 1973-74 oil embargo. Our comfortable, energy-rich world was suddenly powerless. We waited in gas lines, reviling the oil companies. We cursed the Arabs and demanded that the government "do something." It was a dramatic start to the current energy crisis, which has now stretched over two decades.

We believed when it started that it was a true crisis, and the reaction was typical of many crisis periods. There was a great deal of thrashing around in all directions looking for a way to end it. There were attempts to place blame and point fingers. Some people were demanding that we send in the army to take the oil from the Middle East by force, thinking "we have a right to it." It was not a pleasant time, and our response as a people was not one of our finest hours.

It was a crisis that was inevitable—the only question was when and how it would occur. From the moment in 1901 when Spindletop brought the promise of vast quantities of cheap oil to the United States, our economy had been built on the use of energy to expand our productivity and enhance our daily lives. During the early years of the century the US was the largest oil producer in the world. Production supplied all the US demand, and oil was exported to other nations. We did not heed the first warning sign in 1948, when the United States changed from an oil-exporting nation to a net oil-importing nation. The oil reserves, which were so large at first, were proving to be finite in nature and we were using them at a prodigious rate.

When we first began to import oil nobody was concerned. Most of it came from the Middle East and was supplied by American-owned companies at a cost of less than $2 a barrel. The next warning sign came in 1951 when Mohammad Mossadeq, the leader of the Nationalist Front in Iran, gained control and became prime minister. Upon assuming office, he nationalized Iran's oil. When the US realized that Mossadeq and Iran were capable of exploiting the oil resources without US participation they turned to the Shah, the leader of the antinationalist forces in Iran. With direct CIA aid, a military coup ousted the nationalists and restored the Shah to power in August of 1953, thus attaining a more favorable oil policy. The seeds of trouble had been planted, however.

Ten years later in March of 1963 the next big milestone was reached as the Arabian American Oil Company (ARAMCO) agreed to relinquish 250,000 square miles of oil concessions back to Saudi Arabia. This was followed by the nationalization of oil in Iraq in 1972. The stage was now set for the crisis to begin.

At the beginning, the most staggering of all effects, besides the shortage of gasoline, was the rising cost of petroleum products. In 1973, gasoline could be bought for 35 cents per gallon, and a station attendant checked the oil and cleaned the windshield. A few years later, $1.40 per gallon was common, and we filled the tank and washed the windows ourselves. The cost of heating oil took an even deeper bite as we were forced to lower the thermostats in our homes and make sacrifices in other essential areas.

During the 1970s the process of adjusting to the higher real costs of living meant that we could not buy or do as many things as we once could. This resulted in reduced sales of goods and services. Business slumped and people lost their jobs, so now there were fewer people who could buy goods and services. This cycle, once started by the driving force of high-cost energy, continued until a new lower level of balance was achieved.

The economic recession that resulted impacted most of the world. Unemployment was very high. The banking system was balancing on the brink of disaster because of massive loans to third-world nations. Many of these loans were made to help compensate for the high cost of energy. Some were made to oil-producing nations to provide for industrial expansion. Even many of the oil-producing nations were in serious economic trouble as the demand for oil decreased significantly due to its high cost. They had borrowed heavily on the promise of future riches, believing their "black gold" could be sold at any price. However, with the demand reduced, the OPEC nations could not discipline themselves into reducing production enough to sustain the unrealistically high price, and it fell.

The price demanded by the Middle East oil-producing nations had nothing to do with their cost of producing the oil. When the embargo started in 1973, the *cost* of producing oil in the Middle East was approximately *25 cents per barrel.* The *price* was raised to over *$30 a barrel.* We all breathed a sigh of relief when the price fell, but it never returned to its original level.

Meanwhile, the economy struggled back to reasonable health, but our world had changed. We switched to smaller cars, many built by the Japanese. We recognized the need for greater energy efficiency. Industries that were energy inefficient were in serious financial trouble. We were more aware of the finite supply of available oil and how difficult it was to switch to other sources. Very few new oil reserves had been found. No new large-scale energy sources had emerged to replace our dependence on oil. The debt of third-world nations was out of hand. Balances in international economics had changed. Japan had emerged as a dominant industrial nation because they were more prepared to absorb the energy cost increases than the US. Lacking their own resources, the Japanese have had to import energy from the start of modern times—their economy was already geared to very high energy efficiency.

Easing Crisis Brings a False Sense of Security

By the time the decade of the 70s was over, the initiatives made by the US to increase energy efficiency were starting to pay off. The reduced demand for foreign oil helped stabilize the price. Most people thought the crisis was over and forgot about doing anything about it. As the 1980s progressed the usage began to grow again—we had eliminated inefficiencies, but the population continued to expand.

Then in 1990 Iraq seized Kuwait and put a strangle-hold on the oil-hungry world. We were held hostage by a single madman. The price of oil once more jumped sky high. The US lead the combined military forces of many nations to free Kuwait. The banner under which the forces fought was "civil liberties of Kuwait," but we had really gone to war over oil. Fortunately, it was a war that had the total commitment of the United States and the support of most of the rest of the world. Its limited objectives were quickly and decisively achieved. The price of oil dropped again, but not as low as it had been before the war with Iraq. The US was left with one of the longest recessions since the Great Depression of the 1930s.

We were not alone in the recession. Most of the other industrialized countries were also involved. The European Community,

New Zealand, and Australia are all suffering from seriously depressed economies and unemployment. In Japan, where economic growth had been nearly continuous for decades, real estate values dropped dramatically, the stock market tumbled, and their vaunted manufacturing industries floundered—this time, they did not hold the energy efficiency edge.

Japan's story during the oil crisis of the 1970s is an interesting one to review because it was during that decade that Japan emerged as one of the economic and industrial giants. Theories about why they were so successful have been analyzed and discussed for years, focusing on factors such as their management systems, their dedication to quality, government/industry partnership, emphasis on research, and the willingness of the work force to work long hours at low wages. All of these factors undoubtedly did contribute to their success, but one feature that has not been analyzed is the impact of world energy costs and how they opened the door to success for Japan.

Japan has practically no energy resources. They have had to import essentially all of their energy used during this century. Out of necessity, they built their industries around highly energy efficient approaches. When the cost of energy suddenly rose, Japan had to pay the same higher prices as did the other oil-importing nations, but Japan's industry was impacted less because they were *already* using the newest and most energy efficient systems available. Their entire automobile industry was based on small, fuel-efficient cars.

The United States on the other hand was caught totally unprepared. We had recently extracted ourselves from the quagmire of Vietnam. The scandal of Watergate had drained the national confidence even further. Our industries had not modernized to take advantage of more energy efficient processes because there was no apparent need—oil was still cheap. So when the OPEC oil embargo came, our industry was cut off at the pockets. American steel companies could no longer effectively compete in the world market because of their high manufacturing costs due mainly to obsolete, energy-intensive processes. The American automobile industry was still building big, low-mileage muscle cars that very few people could afford to drive anymore. The rest of American industry had similar, if less critical, problems facing them with the increases in energy costs.

In this environment Japan was ready to exploit its advantage in the world marketplace. It took the United States automobile industry nearly ten years to turn around and start to be competitive in the fuel-efficient car market. The steel industry never did effectively recover. During this period of explosive growth, Japanese industry became dominant in much of the world market. Their entire manufacturing industry benefited because of the great influx of capital that flowed into their country.

During that period, the US industry struggled to adjust to the altered energy price pressures and gradually increased their efficiency. As a result the situation has changed in the ensuing years, and the recent energy price increases have affected most of the industrial countries in a similar fashion. Japan no longer has a big advantage, and they are suffering a recession and wondering what is happening to them.

Meanwhile another frightening milestone was reached in the US in 1992. For the first time in our history the number of jobs in the manufacturing industries fell below the number of jobs in government services. There were 18,300,000 workers in the manufacturing industries, which is a primary part of our society that enhances the total wealth of the country. That number is now exceeded by the 18,400,000 people who are working in government services at all levels. This segment of our society simply stirs the wealth around and makes no contribution of its own. It is another very serious sign of the deterioration of our economy.

Since the breakup of the Soviet Union in 1991 and the removal of price controls in 1992, the people of Russia have faced price increases that are far beyond their ability to pay as they struggle to make the transition to a market economy. Their country is essentially bankrupt, and they are looking for help from the West. Many of their scientists and engineers look forward to a jobless future as their efforts to develop machines of war are no longer needed. In addition, it appears that Russia will soon become an oil importer instead of an exporter.

The OPEC nations are gathering their strength again and plan new production limits and price increases. It is only a question of time until the next phase of the energy crisis occurs. It could be very damaging indeed. Conservation efforts have already forced the removal of most of our energy inefficiencies. The next time, we will have to reduce our standard of living dramatically and suffer the resulting economic penalties.

Very little new oil has been found over the last twenty years, and very few new energy sources have been brought on line to contribute to our energy future. In the meantime, most of the nations of the world are in economic recessions. The US is doing little as a nation to solve our energy crisis—aside from closing our eyes to it and planning ways to increase our conservation efforts. Each day we delay brings the day of reckoning one day nearer.

Inflation—Energy Costs in the Driver's Seat

Economists talk of controlling the money supply in order to control inflation. This idea is undoubtedly partially correct, but other factors can also have significant effects. As the real cost of energy rises higher in relation to other costs, it then affects the cost of everything we do. It not only influences the cost of fuel for our cars and furnaces, but also the cost of food and other products we buy.

Looking at our food chain illustrates this point. Our food comes from farms operated by less than 4% of our population. This is made possible by highly mechanized operations that use energy to replace human labor. Many farms are irrigated by water sprinkler systems that use electric power to pump the water. The water flows through pipes made of aluminum manufactured with the aid of electricity. The fields are fertilized with petroleum-based products that require additional energy to process. Once the food is grown and harvested, it has to be transported to market in a system that uses energy to power either trucks or trains. Often the food must be refrigerated to keep it fresh for our tables—another use of fuel.

As you can see, an increase in energy cost affects the cost of food at every step. After increasing efficiency, any further effort to reduce energy will result in decreased productivity, which increases cost. The end result is that the farmers and the processors have no choice but to pass the cost on to us if they are to stay in business. This is one of the hard-core foundations of inflation.

This same sequence of events occurs with every product we use. It often starts with the mining of ore, which is the basis of many of our raw materials. The machines to extract the ore from the ground, whether in deep shaft mines or surface strip-mines, are operated by energy. Next the ore must be processed into raw materials such as steel, glass, copper, and aluminum. The processing normally requires high temperatures, which must be provided by some type of energy source. This processed material is then formed into a useful product—requiring more energy to shape and form. Then comes the finishing—baked enamel, ceramic coatings, tempered glass, or polished brass—all requiring energy. Finally, these products have to be transported to the marketplace—often to an intermediary destination before finally being transported to their place of use. Even though we are now recycling many of these raw materials, much of the cycle must be repeated again.

It is interesting to note that one fourth to one half of the operational cost of most commercial transportation systems is in fuel costs. Therefore, any change in the cost of energy has a large direct impact. Once a product reaches the marketplace—whether in the small corner grocery store or at the mall—it is displayed and stored in heated or refrigerated areas requiring energy. Then, of course, it has to be transported by the consumer to its final place of use, and when its usefulness has ended, the remains have to be transported either to the landfill or the recycling plant.

In many of the third-world nations I have visited, the shop owners cannot afford the cost of lighting or heating their shops. They only turn on the lights if you are going to buy something or, as is often the case, leave the lights off and just let you struggle to find what you want in the dim light seeping through an open door.

During the dramatic energy price increases of the 70s our country experienced serious inflation driven by energy costs. We will experience that again as the crisis progresses and energy costs surge. Will we be like the third-world shop owners and sit in our dark and cold buildings wondering what went wrong?

Communication Sharpens World Energy Demand

One of the technological developments that has become an integral part of our lives is the dramatic expansion of international communications. Its impact on the people of the world is very far reaching. Today, news is transmitted in real-time throughout the world. Televisions and VCRs have provided nearly everyone in the world an insight into what is going on in other places and countries.

Of particular importance is the display of the high standard of living achieved by industrialized free nations. One of the reasons for the failure of communism can be attributed to this opening in communications. People of the former communist countries were able to see how the people in developed democracies lived and began to ask, "What is wrong with our system; why can't we have those things, too?" As these countries evolve into market economies, their standards of living will rise and the demand for energy will increase.

The same expansion of communications has filtered into the underdeveloped nations and exposed their people to new ideas. As I visited many underdeveloped countries scattered across the Pacific Ocean and came to know the people, I found that they all yearn for the things they have seen in the media. Never underestimate the far-reaching power of communication. While they appear to live simple, idyllic lives in their tropical island paradise, in reality their existence is at the subsistence level in most cases. When they are exposed to what they do not have, they become restless. People are the same everywhere, regardless of who they are or where they live. They want to improve their situation in life. Increased exposure to how it could be will bring an ever-increasing demand for development, which in turn is going to expand the need for energy to make it possible.

The Waning Years of the Third Era

We are looking into a future that has a growing world demand for energy. Demands from the newly free, former communist countries. Demands from the underdeveloped nations seeking their share. Pressure from expanding world populations, and the always present desire to improve our own standard of living in the US.

Since the start of the energy crisis in 1973 when the United States was at its economic peak, it has gone from the largest creditor nation to the largest debtor, now owing $1 trillion to foreign countries. United States manufacturers lost domestic market share in all 26 basic industries. Real income for the average American worker has dropped 8% and the per-capita gross domestic product has gone from being the highest to being the 10th highest. The national debt has ballooned to $5 trillion and is growing at the rate of half a billion dollars *per day*.

We are experiencing many similarities to the crisis that England faced when they were running out of wood. This time we are witnessing the beginning of the end of the third era of energy—the Era of Oil. As it was after the end of the Eras of Wood and Coal—wood and coal still exist as fuels today—the same will be true of oil, and it will be used far into the future. However, its days as the major world energy source and price setter are numbered.

The situation is even more serious than just the future of oil. In a paper presented in Dallas at the 16th Annual North American Conference, United States Association for Energy Economics and International Association for Energy Economics, in November of 1994, Richard C. Duncan revealed a startling analysis. His paper reviewed the resource predictions made by the eminent geologist and geophysicist M. King Hubbert over the last forty years. He explored the accuracy of Hubbert's predictions on oil and natural gas compared to actual experience and concluded that Hubbert's 1956 predictions for world oil and gas production are correct. Duncan's further analysis of the data projects that world oil production peaked in 1984 and world gas production peaked in 1990. Duncan's computed analysis based on historical data through 1993 places the peak of world coal production in 2007 compared to Hubbert's projection of 2156. The difference is that Duncan believes that Hubbert's prediction is erroneously based on the amount of coal in the earth, not on the amount that is practical to extract, which is only one tenth of the coal in the ground. If this is true then the world's fossil fuel energy production will peak in 1994 and we are already beyond the point of no return for a smooth transition to a new energy source. To quote Richard C. Duncan, "This means that the world industrial civilization *based on fossil fuels* is now in irreversible decline. There is neither the time nor the energy, *primary energy*, to make a smooth transition to a high, or even a 'sufficiency,' steady-state."

According to Cambridge Energy Research Associates, the emerging economies in Asia are escalating their demand for oil at a rate that will outstrip North America in energy consumption by 1999, while consumption in the US, Europe, and Japan has crept back to its highest level since 1979. Daniel Yergin, president of Cambridge Energy Research Associates, predicts that worldwide daily demand will rise by 11 million barrels over the next decade, an amount equivalent to the combined daily production of OPEC.

Yergin warns: "There will almost certainly be at least one major energy surprise before the decade is out."

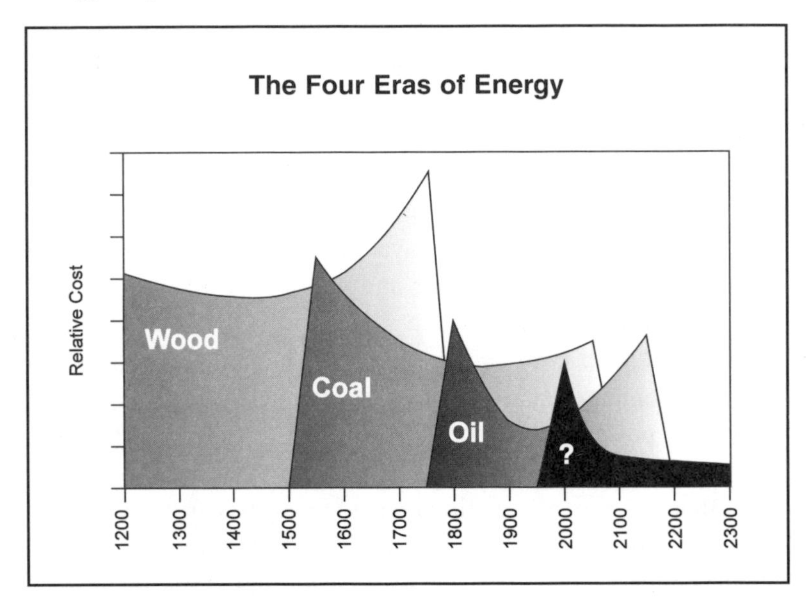

The crisis that started with a traumatic oil embargo is now in full swing, but everyone thinks it is over because oil prices are low at the moment. It does not change the facts of what is happening. How long we wait to find a replacement is up to us. We can simply accept the situation as it is and hope something happens as our comfortable world disintegrates around us, or we can aggressively develop a new nondepletable, low-cost, environmentally clean energy source—a global solution for the coming energy crisis.

5

Criteria For A New Energy Source

In the early 1970s, the United States was riding the wave of abundant wealth—wealth provided by low-cost energy. Then came the OPEC oil embargo. The resulting energy crisis is now two decades old. America has fallen from its economic pinnacle and has suffered the woes of recession. Real income has dropped as the United States has gone from being the largest creditor nation to the largest debtor nation. It has lost domestic market share in all 26 major industries, per-capita gross domestic product has dropped from first to tenth position, and the national debt has soared to $5 trillion.

Ronald E. Bates of the *Chicago Tribune* vividly characterized the situation as "a war of economic survival . . . and the US is losing."

At the start of the 1970s energy crisis, after the first pointing of fingers and blame casting, there were some bold statements made by our leaders that we would strive for "energy independence." Unfortunately, they neglected to tell us how. Jimmy Carter formed the Department of Energy with a mandate to develop new energy systems. The DOE funded several programs including Boeing's solar power satellite studies under my leadership and encouraged other organizations to develop alternative energy sources. However, the emphasis was on systems that could be developed in a hurry and that did not require much capital investment.

While I was briefing one of the senior DOE managers on the solar satellite program, I asked him, "What criteria are you going to use to select the best energy systems to develop?" He looked at me with a blank stare and then replied, "We'll know when the time comes."

I was appalled. If the requirements of a system aren't established at the beginning of a program, there is no way to measure the program's success. The Department of Energy was trying to lead the nation in the development of new energy systems and its staff had not established any criteria except that they wanted the systems in a hurry and they wanted them to be as inexpensive as possible. Not surprisingly, twenty years later, we have still to implement a new energy system to replace the dwindling oil supplies.

Historically, the US has always placed the emphasis on short-term solutions. This policy is a result of our bottom-line, quick-profit–oriented economic focus. Other nations, like Japan, focus on long-term results and are thus able to invest the necessary time and research to devise better products and systems. We see this long-term philosophy paying off for Japan in the form of superior televisions and cars. And, most significant to our current topic, *Japan has **already** performed the first in-space test of wireless energy transmission.* They plan to test a 10 megawatt solar power satellite in low earth orbit by the year 2000 and to provide 30% of the global energy needs by the year 2040. A very ambitious goal indeed and one that if accomplished would place them in absolute economic control of the earth.

After the oil embargo, while the DOE shifted its focus from energy program development to watchdogging the nuclear power industry, the government placed its energy solution emphasis on conservation as the only readily available option. Conservation is easy because it happens naturally; like the default settings on a computer program, it does not require any deliberate actions. Conservation is easily implemented by the force of economic pressures, even if we do nothing else to impose it.

Since oil is the major world energy source, it controls the pricing structure of all the other competing sources. As oil prices rise, the demand for competing low-cost sources skyrockets and their capacity is soon exceeded. The result is an increase in the demand for oil, and the producers can set their prices arbitrarily high in response to the demand.

As the cost of our energy increases we individually find ways to use less in order to balance our own budgets. Oil prices increased fourfold in 1974, and then in 1979, after the fall of the Shah in Iran, they increased another two and a half times. As prices increased, we automatically turned to conservation simply because the increased cost to maintain our previous usage was beyond our means to pay. I joined many others when I traded my gas-thirsty Pontiac Firebird for a Volkswagen and cut down on extra trips.

One day in the late 1970s, after most of the conservation steps had been authorized by the government, I gathered a group of energy experts in my office to help me develop material for a radio debate on the subject of energy development in the United States. I asked them how they would characterize conservation. A discussion followed and finally one of the them said, "Conservation is really the organizing of scarcity and the distributing of less and less to more and more." Another added, "Conservation is a slow walk down a dead-end street."

Conservation has eliminated much waste and has worked well enough to keep the country going, but it cannot provide the foundation for increased productivity or for an increase in our standard of living, nor can it provide the economic growth for which most of us strive, and it can't replace the depleted oil and coal supplies. Unfortunately, the government is still using conservation as its primary energy policy and is not vigorously pursuing alternative energy sources to replace oil.

What Are the Requirements?

Worried about the future of the planet, six Nobel Prize winning scientists were joined by 95 other Nobel laureates and 1,479 scientists from 70 nations in November 1992 to issue "A Warning To Humanity." The statement says, in part, that "a great change in our stewardship of the earth and the life on it is required if vast human misery is to be avoided and our global home on this planet is not to be irretrievably mutilated." Among the actions called for in the statement is to reduce the use of fossil fuels and increase the use of solar energy, wind power, and other energy sources that are inexhaustible and more environmentally benign.

If we are to reverse the recent process of degradation and find a source of energy that can move us from the waning days of the third era of energy into the fourth era, and into the next century, we must find a way to identify the most workable energy solution.

History has shown us that energy is the engine that drives modern civilization. The cost of energy, how plentiful it is, what form it has, and the extent of its environmental impact overshadows all other developmental influences on commerce and industry. We need to find a new energy source that will allow us to regain control over our destiny and move us into the fourth era of energy. The question is, how do we know which energy source to develop to accomplish that goal?

In order to answer that question, it will be helpful to review the reasons that a replacement energy system has not been developed despite the obvious need for one.

The solar power satellite program received very positive reviews when it was first introduced, but implementation threatened the profits of the established energy industry. The nuclear fusion advocates aggressively and successfully lobbied to stop any further solar power satellite energy research. Over the past four decades, hundreds of millions of dollars each year have been spent on fusion research with limited success. The fusion people knew they had to kill the satellite program or they would quite possibly lose their funding. They set out to stifle the one program they knew could be successful in order to maintain their own special program status and income.

I asked several members of Congress how they could justify funding fusion energy development with hundreds of millions of dollars a year when it showed such little progress for the investment. At the same time, Congress was giving solar power satellite development less than ten million a year, and the progress was spectacular. I discovered that, during its early development, fusion research was often commingled with weapons research and no one outside of the classified programs was really sure of what was going on. Technology breakthroughs were often announced or implied, with predictions that usable fusion power was "only twenty years away." In the 1970s, the researchers were still saying it was twenty years away but no one could postulate a realistic design for a working power plant. The damage the fusion lobbyists have caused the American economy is appalling.

Many other systems, such as breeder reactors and synthetic fuels, were also investigated in the 1970s. These systems were being supported by the resources of the nuclear power industry and by the oil industry, but still, they were unsuccessful.

There are two fundamental reasons. After the election of Ronald Reagan as President, the government energy development policy was essentially eliminated. The new administration decided to let market pressures and private industry set price and policy. That put the existing private energy industry in control so they no longer had any reason to fund new energy sources that would compete with themselves. In addition, developing an enormous new energy system is so costly and the time it takes to bring it on line is so long, it is unlikely that private industry alone could do the job, even if it wanted to. However, before Reagan, the government had been successful at funding, developing, and coordinating huge projects that private industry couldn't afford like the national railroad system in the nineteenth century, the highway system under Eisenhower, and the space program under Kennedy.

The government has never followed through on developing realistic requirements for an energy system, and it has totally ignored the natural forces of the marketplace. The focus has been on short-term solutions with low initial cost, ignoring the fact that new energy systems and industries require time and money to develop the necessary technology. Without realistic criteria and a commitment over the long term, there's no way to evaluate which energy systems should be developed. The situation suited the energy industry perfectly as their lobbyists could promote solutions that would enhance their ability to make money without having to meet any requirements.

A case in point is synthetic fuels that are made from coal, oil shale, or tar sands. Due to a lack of oil resources, Germany used synthetic fuels in the second World War and South Africa used them when embargoes threatened their oil imports. Both the extraction and burning processes of synthetic fuels produce high levels of pollution, and the extraction and refining costs are astronomical. Congress, during the Carter Administration, budgeted $88 billion for synthetic fuel development for the US. When measured against any set of realistic criteria, synthetic fuels have no long-range merit except to make their developers rich. They are fuels of last resort, not energy to run the future.

Establishing criteria to fit our needs is logically the first step for solving any problem. If you don't define what you are trying to accomplish, there's no way to measure your success. Properly selected criteria should define the needs of the goal without implying or defining a specific solution. The Department of Energy's mistake was to define a solution instead of a need. That approach has prevented the serious evaluation of alternative energy sources. The opposite occurs when a need is defined by a set of criteria or requirements—the possibilities are expanded.

Setting criteria is a little like setting rules for a game. Before the game starts, all of the participants must agree on the rules that will measure how each performs. In the same way, participants in a new project must agree on the criteria before starting the development phase so the results can be evaluated the same way by all the participants.

Criteria represent the goals that we want to achieve. When establishing criteria, there are three questions that need to be asked: *What is it we are trying to accomplish? What is wrong with the way things are now? What specific changes would improve the situation?* Defining criteria is the process of putting goals into quantifiable form so that we can measure our progress towards the goals.

When a requirement is established that is beyond currently existing technology, the challenge to the engineers usually results in new and innovative solutions. It opens up new areas of technology to meet the requirements. For example, if a requirement for a new airplane is the ability to carry a heavy payload a longer distancc than has been previously possible without refueling, then the engineers have to find ways to reduce weight or increase fuel efficiency, or develop some other idea in order to meet the criteria. Examples of technology developed in response to need include infrared vision which was invented in response to the need for weather satellites to be able to film the earth through the clouds, and the advanced jet instrument technology invented when planes first exceeded the speed of sound, needed because the vibration made the instruments at the time inaccurate.

If we are to identify an energy system that can replace oil, we have to start with the right criteria. When we started the solar power satellite project in the 1970s, one of our first tasks was to develop a set of criteria for a new energy system so we could measure how well we were progressing and have a way to evaluate whether the concept would be successful. I tried repeatedly to encourage the Department of Energy to use the criteria, but was never successful. The DOE was staffed mainly with nuclear advocates, and one of their primary functions was to develop nuclear weapons. It was not surprising they rejected our energy criteria because it would not have supported the continued development of advanced nuclear systems.

As a free and idealistic nation, there is no reason not to include criteria for a new energy system that reflect our values. These include our desire for a higher standard of living for ourselves and for the rest of the world. We can take into account that our world is a finite globe that must be protected if it is to provide a home for all of its creatures in the future. Our criteria can recognize that we don't believe any peoples should be oppressed for the benefit of others. Applying our values to the criteria means that we must look toward a low-cost energy source so that it can be equally available to everyone.

What gives us so much confidence in solar power satellites as a viable energy option for the future is the fact that they can satisfy the criteria, including those that reflect our values. In the talks and briefings I gave about the solar program, I always explained how the system satisfies those fundamental requirements. The public understood perfectly why they were so important. Unfortunately, like all government agencies, the Department of Energy simply could not accept the idea of losing control by accepting criteria that would allow independent solutions.

The government still has not established realistic criteria for an energy solution for the future; for an energy system that is suitable for the fourth energy era. Since the Department of Energy has failed to perform this most fundamental function of government, I will advance the set of criteria we developed when we first started the program at Boeing. They are as valid today as in the 1970s, and they are absolutely critical to the selection of a new energy system. The reason there is no viable new energy system being developed today is because of the failure to establish valid criteria for making decisions. Instead the effort has been and is still being applied to programs and systems that can never satisfy the needs of the future, but are the pet scheme of some bureaucrat or politician. Enormous sums of money have been wasted, when instead they could have been applied to solving the problem if only the decision-makers had used a few simple criteria to measure the value of the product they were trying to develop.

How Much Can it Cost?

The first energy crisis occurred in England in the sixteenth century when the demand for wood exceeded the supply. Those who could afford it paid exorbitant prices to prospering wood importers while the poor did without. In response to the dwindling fuel supplies, the large coal deposits of England were mined and substituted for wood. Not only was coal a satisfactory substitute, it also offered many advantages. It was abundant, low in cost to mine and transport, and relatively compact so it provided more energy for the weight than an equivalent amount of wood. Coal also burns hotter than wood and the increased energy could accomplish more tasks. Because of its vast coal supplies, England rose to a position of industrial and economic dominance in the world in the eighteenth century, despite its limited land area. England maintained its leading economic position in the world until the United States—floating on a sea of oil—burst on the scene.

Oil was first used extensively during the latter part of the eighteenth century. Its use was limited at first because of the high price of extraction; coal still dominated the energy market. As the availability of oil expanded, the price dropped and usage increased. Whale oil, which had been used for years as lamp fuel, was replaced by petroleum products because of the lower cost. When vast quantities of oil were discovered in the US after the turn of the twentieth century, the price of oil dropped and use expanded dramatically. The productive capacity and economic standing of the United States multiplied until, by the end of the second World War, its productivity had no equal on earth. Our expansion had been built on the availability of low-cost energy. For decades the US remained the dominant world economic power because its oil continued to flow out of the ground like water. Early on, even the oil that came from the Middle East was pumped by US companies who paid only a small percentage of the profits to the source country.

History has shown that, wherever low-cost energy appears, economic prosperity follows. Though low cost cannot be the sole criterion for an energy source, the cost must be in an acceptable range that will make the power widely available or no large-scale development can occur.

Energy cost does not necessarily have to be low during the development stages, but it must be low over the long term. In the early part of the eighteenth century when oil was first being developed as a power source, it was selling for two dollars a gallon—extremely expensive by the standards of the day. By the 1930s, the price of oil had dropped to twenty-five cents a barrel nationwide, and even dropped as low as two cents a barrel in West Texas.

Therefore, the first and most obvious criterion for an energy source is *low cost* after the development stage. However, there are other, more important considerations if an energy source is to have longevity and keep up with an increasing world population.

How Much Energy Do We Need?

Each succeeding era of energy passes at an accelerating pace as the demand for energy grows with an increasing world population and expanding industrialization. The first energy era, the Era of Wood, came to an end because, even though wood is a renewable resource, it could not be grown fast enough to satisfy demand. Today, impoverished nations are still stripping their lands of wood to cook food, keep warm, or to sell for hard currency.

During the second and third energy periods we used stored energy in the form of coal and oil. It took millions of years for these fuels to accumulate in the earth, and we are devouring them in mere centuries with only a fraction of the earth's population participating in the feast. There are literally billions of people in the world looking toward the industrialized nations with envy and hope as they struggle to achieve the level of prosperity we have enjoyed because of our energy resources.

Oil in the United States seemed limitless after the great oil fields of Texas were opened. Americans could afford to use any amount they wished. Oil created the giant industrial nation of the twentieth century, but after only fifty years we have had to resort to imported oil to satisfy our desires. There are still large reserves of oil in the world, but the end is in sight after less than a century into the third energy era.

The United States alone consumes a quarter of the world's energy. Maybe we are using more than our share; perhaps we should cut our consumption. But we have already felt the results of reduced energy use since the start of the oil crisis. Our standard of living has fallen and our economy has stumbled badly. Conservation, which has dramatically increased efficiency in energy usage, has also contributed to our lowered standard of living. More importantly, increased fuel costs were a primary cause of our economic decline. Industries moved abroad, unemployment rose, and the national debt has exploded. Our high standard of living was based on the availability of ample, low-cost energy. A further reduction in our energy usage will eventually reduce us to the poverty of the third-world nations.

What about the rest of the world? The first time I visited China in 1979, it was just beginning to open its borders to increased trade with the outside world. By the time of my second visit in 1985, it had made great strides in modernizing its society and developing a market-based economy. Today China has one of the fastest growing economies in the world. Its mission is to raise the standard of living to match that of the Western world. One of the first goals is to provide each family with a refrigerator. After walking the streets of the Chinese cities, surrounded by hordes of people, it is hard to imagine the number of refrigerators that would require. The task of manufacturing and distributing millions, if not billions, of refrigerators requires a huge amount of energy—enough to increase world energy consumption by 7%. If China achieves its long-range goal of matching the standard of living of the United States, world energy consumption would double. How long will the oil last and how high will its price go if the available energy has to supply China's increased demand? And China is only one nation. To raise the rest of the world's underdeveloped nations up to our standard of living would again double the amount of energy needed. But we can't deny the rest of the world the opportunity to advance.

With the energy demand increasing throughout the world, we can no longer rely on finite resources stored in the earth. Therefore, as a second criterion we need a *nondepletable* energy source that is either so vast that we cannot consume it, or a source that will renew itself at a sufficiently high rate to keep up with demand.

We Can No Longer Ignore the Environment

We are learning—from sad experience—that our environment is far more fragile than we had imagined. The earth cannot tolerate the abuse of billions of people without the environment changing in ways that may be hostile to human life. Plant and animal life forms are disappearing at a rate that is much faster than the natural course of evolution, and unless we cease our carelessness, we will surely destroy what is now remaining, and ultimately ourselves in the process.

Burning wood, coal, and oil has sharply increased the amount of carbon dioxide in the atmosphere with devastating effects. The increase is causing negative changes in the world climate and threatening the earth with the possibility of a runaway greenhouse effect. Carbon dioxide levels have been increasing at an accelerating rate every year since they were first recorded in 1957.

In the 1950s, London was plagued by "killer smog" caused by burning coal used to heat homes. The smog of Los Angeles is known worldwide, but less well known is the greater smog caused by coal and charcoal in the cities of China. When the Eastern Block nations opened their borders, the world was given a glimpse of the horror that careless burning of fossil fuels can cause. In the Romanian town of Copsa Mica, soot from coal burning has stained everything black; it is known as "black town" where animals die within a few years inside the city limits. In the former Soviet Union, the Ukraine produces eight times as many atmospheric pollution particles as the entire United States. In Canada and the United States, acid rains caused by coal burning are damaging mountain lakes and forests. In Germany, the Black Forest is dying from coal pollution.

In areas around the world where wood is still available, wood-burning stoves and fireplaces are used to save money on heating bills, but wood smoke degrades air quality to dangerous levels. In the United States, wood-burning bans are now common during the winter months.

Both fossil fuel and nuclear power plants generate thermal pollution in addition to electricity. In the process of converting heat energy into electricity, two thirds of the heat is wasted and must be absorbed into the environment, but even one extra degree of heat can render a lake deadly to the fish that live in it.

Other forms of nuclear contamination are even deadlier than thermal pollution. The Chernobyl nuclear reactor disaster spread radiation contamination over vast areas of the former Soviet Union and its neighboring countries. Even wood from the area can't be burned or milled into products because of the radiation residue in the trees. There is also the matter of the radioactive waste that accumulates as a by-product of generating electricity from nuclear power. The cost of cleaning up nuclear waste is never included in the cost of nuclear power generation, but it is a cost born by all taxpayers nevertheless. An example is the cost of cleaning up the waste at the Hanford nuclear facility alone—in the tens of billions of dollars. No one has yet answered the question of how to safely store or dispose of the waste.

All of our current energy sources damage our environment. They pollute or eliminate our food sources and poison the air we must breathe to stay alive. If we don't find a way to restore the environmental equilibrium, we will perish. But as the underdeveloped nations raise their living standards, world energy consumption rises. We need to develop an energy source that can exist within our environment without destroying it. Therefore, the third criterion is an *environmentally clean* energy source.

Availability Equals Opportunity

When oil began gushing from the Spindletop well in Texas in 1901, it heralded the large-scale expansion of our economy. We were one of the few industrialized nations able to reap the maximum benefits from oil because we were one of the few with sufficient internal supplies of oil and natural gas to meet all the energy demands of the first half of the century.

Today, most industrialized nations, including the United States, rely heavily on imported energy for their economic viability. For the past twenty years, the United States has experienced an economic downturn resulting from its oil sources being controlled by other nations. Other countries experienced the problems of foreign oil control long before we did. Japan, for example, has no internal reserves and has to import nearly all of its energy. A desire to have its own oil source was one of the main factors in Japan's decision to enter into the Second World War. The attack on Pearl Harbor was aimed at destroying the United States' Pacific fleet so Japan could control the shipping lanes between Indonesia and Japan after they seized the oil-rich islands of Indonesia.

Iraq invaded Kuwait to gain control of its oil, as well as to expand Iraq's influence in the Middle East. Control of the Kuwaiti oil supply was also the real reason the United States went to war against Iraq in 1991, as most of us know. The Middle East, with its great reserves of oil, continues to pull the economic strings of much of the rest of the world.

Because of the tremendous leverage energy exerts over the economy of the world, the oil-producing nations are able to wield political and economic influence far beyond what their share of the population warrants. The oil-consuming nations are held hostage to their whims, and tensions run high. As the oil resources of the earth are depleted, the Middle Eastern influence will increase, as will the tension, leading to the probability of new wars.

In order to avoid this continual source of conflict, our new energy source should be one that can be equally available to all nations without regard to location. Therefore, our fourth criterion is an energy source that is *available to everyone.*

Energy Form for the Future

When wood was the world's fuel, it was in an ideally usable form for the time. It grew nearly everywhere. Travelers walking the trails from place to place did not have to burden themselves by carrying fuel for their fires; the fuel was there when they arrived. They just had to gather it together for the evening fire. It not only provided heat, but delighted the eye with the flames and gave comfort with the scent of wood smoke on the evening air.

As the demands for energy grew, wood became impractical. It was too bulky for easy transport and did not burn hotly enough for iron and steel production unless it was first converted to charcoal. Coal, on the other hand, burned very hot and was reasonably compact. It was much easier to feed a steam boiler by shoveling coal than using wood. When ships turned to steam power, coal could be carried in sufficient quantities to drive them across oceans, an impractical trip using wood.

Oil offered many advantages as an energy form. Its various liquid and gaseous states could be readily stored in tanks, pumped through pipes to any location, and could be introduced into an engine in incredibly small quantities—all impossible with coal. Oil seemed to be the ideal form of energy. For many applications it is still the only practical form, but civilization needs more than a liquid fuel.

As the last century evolved we began converting our energy to a higher and more useful form. This higher form of energy is electricity. A large share of our quality of life depends on this form of energy, and the burning of fuel is only one way of generating it. Electricity is energy in a pure form—silent, available on demand, and with no release of pollutants. It is energy that can be directly converted to any service we want. Electricity is the highest form of energy we have.

Electricity provides us with light, telephones, television, radio, computers, kitchen appliances, factory machinery, heating, and cooling—all available at the flick of a switch. Though there are still some things that seem impossible for electricity to power, like aircraft, who knows what new technology will be revealed in the near future? The electric battery technology available right now could replace 70% of our liquid-fueled automobile travel.

What form should the energy of the future take? Should it be another fuel? Can another fuel ever fill the requirement of being nondepletable and nonpolluting? What form will fit the requirements of tomorrow?

Our fifth criterion is that our energy source must be *in a usable form.*

Criteria for the Fourth Era

We now have five energy criteria for the Fourth Era. Energy from our new energy source should be:

1. Low cost
2. Nondepletable
3. Environmentally clean
4. Available to everyone
5. In a usable form

These five requirements cover all of the critical aspects of a future energy system but do not define a specific solution. The test of their validity is in their inherent logic and in the authenticity and comprehensiveness of the data used to develop them. The criteria are derived from what history has taught us about bringing prosperity, preserving the earth, and minimizing world tensions. Each broad criterion addresses a generalized goal. As potential solutions are identified, the criteria can be subdivided into more specific requirements that will help us evaluate how a solution might satisfy the broad, general criteria.

With the criteria defined, it is possible to measure how a known energy source might satisfy the requirements or might point in the direction of a new solution. For example, criterion number two specifies that an energy system must be nondepletable. That rules out all stored fuel because stored fuels are finite by nature and all would eventually be depleted if they were used in large quantities. As a result, we need to consider a system that is either continuously replaced or one that is vast enough to last indefinitely. This exemplifies the process by which the criteria can guide us toward a solution without inhibiting the reasonable possibilities.

The criteria may appear simple, but they make the task of finding a suitable energy source for the future extremely difficult. If a way can be found to satisfy all of them, the benefits to our country and the world will be well worth the effort. An energy system that could satisfy these criteria would form the basis for the fourth energy era. While not entirely replacing all other energy sources, it would provide the core and help keep the price of the supplemental sources within reason.

6

Exploring the Options

The energy crisis initiated by the oil embargo of 1973-74 led to the investigation of many potential alternative energy sources. Some were explored by the energy-generating companies—both privately and publicly owned—and some were proposed by individuals. Others were under the direction of developmental agencies. The federal government combined their efforts within the Department of Energy, which was formed by merging several governmental agencies. During the ensuing years of the 1970s the activity level was very high, but with the breakdown of the oil cartel the urgency was reduced and much of the activity has returned to business as usual. However, a great deal of basic knowledge was generated and many ideas were suggested. Let's review what was considered at the time and what has transpired in the meantime.

Conservation—The Organizing of Scarcity

In the 1970s, conservation was the primary emphasis because it could be immediately effective. The theory is that the fastest and cheapest source of energy is that which is saved. This is certainly true; however, there are secondary effects of conservation that can be devastating if this is carried beyond improved efficiency for any length of time. There are three methods of achieving conservation of energy, and all three have been used extensively.

First and foremost is price. As the cost of energy rises, increased either by natural effects of the marketplace or deliberately, the consumer is forced to use less. This certainly happened to us in the 70s. Oil *prices* (not *costs*) were increased 1000% by the OPEC nations. As price was increased, we automatically turned to conservation; we had very little choice. Some people borrowed money to maintain their previous lifestyle, but that could not go on for long and reality soon caught up with them. They, along with the rest of us, soon found ways to reduce our energy use. We drove much less by turning to carpools, taking fewer Sunday afternoon drives, and using smaller cars with standard transmissions. We closed off the extra rooms and lived in one room instead of the whole house. We added insulation and weather stripping. We turned the thermostat down. These forms of conservation became involuntary when we could no longer afford not to follow them.

The second form of conservation is conservation forced by government decree. In this case, laws are passed or executive orders are issued. Prime examples of forced conservation in the US are the 55-mile-per-hour speed limit, the reduced thermostat settings mandatory in all commercial buildings, the mandatory increases in average gas mileage of all new automobiles, and building code changes that set high insulation standards and limit the area of windows allowed.

Rationing of energy is the next step in forced conservation. Standby rationing was placed on the books in the 70s, but fortunately was not imposed. A form of energy rationing was used in some parts of the country—in some states, gasoline could only be purchased on odd- or even-numbered days, depending on license number. Electric power utilities refused to install services for new houses in some locations because they had insufficient generating capacity to add new customers.

The third form of achieving conservation is to appeal to the patriotic and moral character of the people without committing to any positive action. There have been government media campaigns touting conservation "for the good of our country" and because it is the "right" thing to do. We should insulate our homes. We should not keep our thermostats set at a comfortable level; rather, we should experience some suffering or we are not doing our part. We should not drive alone in our cars; we should carpool or ride buses. We should strictly obey the 55-mile-per-hour speed limit.

Conservation was the only way we could attack the immediate problem in the 70s, but if we rely on this approach as the only permanent solution, we are surely building the major elements of economic disaster and a continuing reduction in our standard of living. Conservation cannot provide the foundation for increased productivity or an increase in our standard of living, nor can it provide the economic growth for which most of us strive. It will destroy the American dream if carried to extremes. Unfortunately, conservation is the primary course we are currently following.

The Resurrection of Coal

The United States has large reserves of coal, and in fact many parts of the country rely on it for electric power generation. Coal supplies more than 27% of our total energy use and 56% of our electricity. It has been a major world fuel for centuries, but it has been replaced by alternative sources whenever possible. There are several reasons for this. It is awkward to obtain. Whether it is mined in underground cavities or in strip mines, the problems are severe. In underground mining the safety and health of the miners is a major concern; the incidence of accidents is near the top of any profession, and the health problems are among the worst known. The annual fatality rate would be unacceptable if we had to deep mine all of our future energy.

Strip mining has its own set of problems. First among them is the damage to our environment and the great areas of the earth that must be devastated. The land areas required are larger than for any other known energy solution.

After coal is removed from the ground it must be burned to provide thermal energy either for direct heating or conversion to electric power. This means the coal must be transported to the geographic locations requiring the power, or the electric power plant needs to be located at the mine and transmission lines built to carry the electricity where it is needed. For example, if the decision was made to transport coal via train to a large power plant with the generating capacity of Grand Coulee Dam, it would take more than one thousand standard coal cars *per day* to maintain the boiler fires of one facility.

Even though there are great reserves of coal in the ground, much of this may not be usable because of the difficulty in mining it. If Richard C. Duncan is correct in his analysis of the worldwide fossil fuel reserves, then only 10% of the known reserves can be economically extracted. In that case we will soon be running short of coal at the current rate of usage.

The other problem with coal, however, is what happens when massive amounts are burned in the earth's atmosphere. History has many graphic examples: Pittsburgh, the city built on coal, with coal, and under coal soot; London, with its coal-smoke-augmented smog that ultimately resulted in many fatalities, corrected only when the burning of coal in private homes was prohibited; China, with few private automobiles, has coal-induced smog in their cities that rivals car-choked Los Angeles at its worst in the 70s; Eastern Europe, where an environmental nightmare was revealed when the borders were opened, with the burning of coal as a major pollutant; Germany, where acid rains in the northeast part of the nation are destroying mountain lakes, injuring crops, and inhibiting forest growth even after most of the visible contaminants have been removed with modern scrubbers required by stringent environmental regulations.

What would it be like if we started burning coal seriously in the US? We are already facing the frightening aspect of the greenhouse effect on world climate due to excessive levels of carbon dioxide in the atmosphere. The world air temperature has already risen more than a degree and a half over the last few years. The natural disaster of the 1991 eruption of Mt. Pinatubo in the Philippines had one favorable side effect for the world: its dust cloud has effectively reduced world air temperatures by about half of the increase caused by carbon dioxide. But this is only a brief respite, and we certainly can't count on a volcanic eruption every few years to counteract the damage we're doing.

The primary effort in expanding the use of coal has been to reduce the impact of its emissions. Much has been done to remove many of the obnoxious exhaust particles, but there is no solution for the carbon dioxide. Each step taken to clean up the combustion products adds cost—and there is a long way to go, so the real cost will continue to escalate. Much of the rest of the world, outside of the US, has done little to minimize atmospheric pollution from coal.

Natural Gas—Oil's Unwelcome By-Product

Natural gas has been oil's partner in the abundant energy world of the twentieth century. Often it was an unwelcome by-product of drilling for oil and was frequently burned off at the well heads just to get rid of it. Today it is becoming increasingly important in the US effort to minimize air pollution from burning fossil fuels. It is one of the cleanest burning fuels available in significant quantities in America. A great number of households across the land depend on natural gas for cooking and heating. Many vehicles are being converted to its use. Consumption of natural gas will undoubtedly expand as costs and environmental impacts from other fossil fuels increase. In the US today nearly all new electrical generating capacity is produced by natural gas turbines. They are cheap to build, and gas prices are currently low. This has lulled the utilities into a false sense of security. Unfortunately, natural gas is a finite resource that, if Duncan's analysis is correct, has already passed its peak of production in the world and will soon join the league of scarcity that will drive its price ever upward.

Natural gas is a very good option for the US as an interim fuel until a new energy system can be developed. It is certainly more desirable from an environmental standpoint than coal.

Nuclear Power—Future Unknown

After the atomic bombs dropped on Hiroshima and Nagasaki brought an abrupt close to World War II, it was only logical that there would be attempts to use atomic energy in many other ways. Its promise seemed unlimited. The amount of energy locked inside the atom boggled our minds. Scientists talked about powering an automobile with a fuel capsule the size of a pea. Electricity would be so cheap it would not be necessary to meter it. Ships would cruise the oceans of the world without the need for refueling.

The euphoria of atomic energy did not last for very long as the difficulties of making the predictions come true were revealed. However, work progressed quite rapidly and by 1951 the first electric power was generated by nuclear energy at Arcon, Idaho. This was followed by the first civilian atomic power plant at Schenectady, New York, in 1955. Atomic power was applied to military ships in 1954 when the US submarine *Nautilus* was converted to nuclear power and the nuclear-powered aircraft carrier *Forrestal* was launched.

These were all successful ventures, and other plants and ships were constructed. There were some serious problems, however. Small power plants were proving to be impractical for most applications, energy conversion efficiency was not very high, and the radiation hazards were a major concern. In the meantime the Soviet Union had developed the atomic bomb, along with England, and other nations were also preparing to join the atomic club. The hydrogen bomb added to the uncertainties of the world.

Even so, many nations were turning to nuclear power plants as a source of electricity. Their use expanded during the decades of the 1960s and 70s.

Even before the accident at Three Mile Island, Pennsylvania, in 1979, fear of the atom had become a fundamental problem in the United States. Serious questions were being asked. Is a nuclear power plant safe? Do we want one in our backyard? Will earthquakes destroy it and subject us to lethal or disfiguring radiation? Some questions were asked out of ignorance and some out of real concern. Emotions ran high on both sides of the issue. Many were motivated by blind fear and the fact that nuclear power burst on the world with two violent and devastating shocks. Two cities were vaporized. Masses of humanity were gone or cruelly maimed. The people of the world were sensitized to the power of the atom.

A very simple analogy can be used. If you have walked on a carpet on a dry winter day and then touched a light switch and experienced the instant jolting shock of static electricity, you will know what I mean. How many times does it take before you can hardly bring yourself to touch the switch again? Yet it is that same basic electricity in a different form and fully controlled that will light the lights in your home when you turn on the switch. Normally, you do not receive a shock. There is no connection between the two phenomena, except that they both involve electrons and are electrical in nature. The atom bomb and an atomic power plant are similarly dissimilar, with the common element being the energy in the atom in this case. Yet we have been sensitized by the devastation of the atom bomb to fear anything connected with atomic energy.

The resulting situation with nuclear power is very confusing. During the late 70s, after the oil embargo, the Carter Administration took opposing approaches to atomic energy. The first policy was to accelerate the licensing of new conventional nuclear plants and pursue research on breeder reactors, but at the same time they banned the construction of new fuel-processing facilities that were needed to extract the remaining useful fuel from used fuel rods. Thus, both the problems of obtaining sufficient low-cost fuel and of storing nuclear waste were compounded. The situation has not improved since then.

Nuclear power supplies about 22% of the electricity in the United States, which is 8% of our total energy use, but the safety issue and nuclear waste have become serious considerations. As a result, new plant construction in the United States has stopped. The uncertainties of nuclear power were highlighted by the accident at the Three Mile Island plant. Then came the disaster in 1986 at the Chernobyl Power Station near Kiev in the Soviet Union, with clouds of fallout raining down on all of Europe in addition to Russia. The uncertainties turned to certainties. The circumstances that caused the Chernobyl accident may not be applicable here, but it happened and nothing can change that in the mind of the public.

People do not understand clearly what they cannot see, feel, touch, or hear, and nuclear radiation fits all of these categories. The public must rely on scientists and engineers for information. Many people doubt the honesty of the highly educated on matters that cannot be readily understood by the average citizen. Unfortunately, in some cases they have good cause to doubt. They have been duped by the intellectual elite from time to time, either through deliberate actions or simple intellectual stupidity. It will take a lot of convincing to make the public comfortable with nuclear power, and in the process of adding the necessary safeguards the costs will be escalated dramatically. An example is the 1,150 megawatt Seabrook reactor in New Hampshire, which was completed in 1986 at a cost of $4.5 billion—four times the original estimate. Much of the price escalation was driven by safety concerns.

One problem that receives little attention is the question of what to do with a nuclear power plant when it is worn out. Because of the temperatures, stresses, and nature of the environment in which the machinery must operate, the plant has a finite life of 30 to 40 years. Some of the early plants are now reaching that age. The Trojan plant in Oregon, on the Columbia River, developed serious maintenance problems and was permanently shut down after only 18 years of operation. The cost of decommissioning the plant has been estimated at $450 million. Should it and the other old plants be rebuilt at enormous cost so they can continue to operate, or must they be decommissioned and their giant cooling towers left to stand as tombstones over their graves?

Some plants have died in infancy. The Satsop nuclear plants in Washington State were abandoned after runaway cost increases due to changing safety requirements and bad management drove WPPSS (Washington Public Power Supply System) to stop work before they were completed—causing WPPSS to default on the bonds used to finance the venture. The plant's cooling towers rise in ghostly solitude above the surrounding forests, mute testimony to the debacle. The only signs of life are the flashing strobe lights warning passing aircraft that here lies one of man's failures. Even the power to illuminate the night has to come from another source.

Many nations, such as France, turned to nuclear power because they felt it was the best option at the time; even today, they look toward it for the future. They either do not have their own fuel—such as oil or coal—or they could foresee the time when those resources would be gone. Even with the increasing costs of nuclear fuel and power plant construction, nuclear power is still cheaper than generating electricity with oil.

One limiting factor for nuclear power on a large scale is the availability and cost of the fuel. One way to extend the uranium is to develop and build breeder reactors that have the potential of extending the useful energy in the fuel by about a hundred times. The problem is generating plutonium, which can be used to make bombs as well as power plant fuel. Extensive research has been carried out in the US, but no operational breeder reactors have been built for power generation.

Many years ago, shortly after I began my career in the aerospace industry at Boeing, we started hearing rumors about a gigantic nuclear explosion in the Soviet Union. There was no mention of it in the news media, but among the engineers the feeling was that something big had happened and nobody was quite sure what. It was not until after the disintegration of the Soviet Union that the truth about the extent of damage caused by Soviet nuclear activities was revealed. In 1957 a nuclear-waste storage tank at Mayak exploded, sending vast quantities of nuclear radiation into the air. Those old rumors were finally confirmed.

The problems at the Mayak plutonium production plant had started much earlier when in 1949 it began spilling radioactive waste directly into Chelyabinsk's Techa River. Today the Chelyabinsk region in the southern Ural Mountains is considered to be one the most irradiated places on the planet. Radioactive waste was also dumped into Lake Karachay. The lake is now so irradiated that just standing on its shore for an hour could be fatal. If there were to be an explosion in the waste storage tanks now at Mayak it could discharge about 20 times the amount of radiation released during the Chernobyl disaster.

The extent of the Chernobyl disaster is also becoming clear as we see on television the families and children suffering the ravages of nuclear radiation exposure. The picture is very grim, and the cost of cleanup is beyond the ability of the bankrupt former USSR. Worse yet is the fact that there are over a dozen other plants of the same design as Chernobyl, leaving these countries with both a current disaster it cannot handle and the possibility of more potential disasters shadowing the future.

In other places spent nuclear fuel has been indiscriminately dumped into the waterways, posing a serious danger of contaminating drinking water. Several nuclear ships have dumped damaged reactors and used nuclear fuel assemblies into the sea. Decommissioned nuclear submarines lay in the port of Murmansk on the Kola Peninsula leaking radiation into the sea and air. Nuclear waste in Russia is a cloud of doom hanging on the horizon.

Disposal of nuclear waste is not just a problem in the former Soviet Union. The United States and other countries using nuclear power are faced with the problem of disposing of the waste, some of which has a half-life of centuries or more. So far no good methods have been found that will ensure its safe disposal for indefinite periods. In the meantime billions of dollars have been spent trying to solve the problem.

The future is very clouded for nuclear power. It has been used for decades, but no clear path has emerged for its expanded use, and time has not been kind to its proponents.

Synthetic Fuels—High-Cost Insurance

Much effort has been applied to extracting what is generally called synthetic fuels. This category of fuels was one of the major thrusts by the government in the late 1970s. Congress approved $88 billion to be spent on developing synthetic fuels. These fuels take many forms, with the most common being processing coal to produce liquid fuels, extracting the oil contained in oil shale, and processing tar sands. All of these sources have been known for many years. As I mentioned earlier, it was the use of kerosene made from coal that stimulated the drilling of the first oil well in the United States.

Several pilot plants were built in the 1970s to refine the processes and determine costs. Unfortunately, the fuel costs were higher than for fuels made from natural oil, and the facilities for processing large quantities of these fuels were not constructed. The costs were being driven by the amount of energy required by the process and other major obstacles to overcome. Massive amounts of materials must be mined and handled. Large quantities of water are required for some of the processes, and as is true with the coal slurry pipelines, the sites of the tar sands and oil shale are generally in areas that have limited water supplies. But most of all, it takes enormous quantities of energy—however generated—to create these fuels that are themselves intended as energy sources.

Germany depended heavily on synthetic fuel made from coal to fuel their war machines in the late stages of World War II. South Africa is currently the only nation with large production capability for making synthetic fuel from coal. It was forced to this method by politically motivated trade embargoes and outside pressures.

Much of the $88 billion appropriated for synthetic fuels in the US was never spent as the true costs of the resulting fuel became known. By then the oil embargo was over. Foreign oil prices had dropped, and the politicians thought it was a good idea to just forget the whole thing. It is doubtful that we could expect this category of fuels ever to be used extensively unless there is no other alternative.

Earth-Based Solar—The First Step Toward a New Future

Probably the most popular of the new initiatives of the 1970s, and one that was enthusiastically pursued by the government, private companies, and individuals, was earth-based solar power. It seemed to fit all the ideals that people could imagine. The source was free, it was nondepletable, it was environmentally clean, it could be utilized by everyone, and it could be distributed in a way that would eliminate the need for depending on a utility company for power. Many thought it was the wave of the future and the path to their personal utopia. It was the showpiece of the Carter Administration's energy development plan, which had a goal of solar energy providing 20% of the nation's energy by the year 2000. How could it fail, as sunlight could deliver the equivalent of 4,690,000 horsepower per square mile of the earth's surface?

A large variety of earth-based solar options were investigated, and several are being used in limited application today, but the early promise has been difficult to achieve for reasons I will discuss as we go along.

Many of the approaches were directed toward heating buildings and water, or using solar cell panels for providing localized electric power. These are generally known as "distributed energy" sources, or "soft technology" energy. In addition to the distributed forms, small-scale electric power plants have been developed. The most significant are windmills (often called wind turbines), thermal systems that use mirrors to concentrate sunlight to heat a working fluid that drives turbo-generators, and various arrangements of solar cell farms that provide electricity directly.

Unfortunately, all ground-solar systems suffer from two basic problems. The first is the intermittent nature of "solar insolation," which is the amount of sunlight that shines on a given area. To put it simply, the sun goes down at night, and clouds often hide it during the day. This usually happens when we need the energy the most—during the winter and when it is stormy. As a result, the second problem arises. In order to make a ground-based solar system work as a complete energy supply system, it must be significantly oversized, generate energy above peak requirements, and be equipped with an energy storage system. Generally, to keep these components from becoming too large, a conventional backup system (i.e., another method of generating energy) is also required.

Solar Heating—The Simplest Form of Solar Energy

The simplest form of solar energy uses sunshine to directly heat buildings or water. We can obtain some advantages from it by simply controlling the window shades on the sunny side of our homes. However, that is not the level of energy usage that can address the serious energy needs of the nation, so I will restrict my discussion to heating systems that could make significant contributions.

The cost of most things we buy varies with size or capacity, and a solar system is no different. Because the energy from the sun is only available for less than a third of the time in the winter, even under ideal weather conditions, the system must be large enough not only to heat the building or water during the time the sun shines, but also at the same time provide enough extra heat to place in storage for the time after the sun has set. It is a little like buying an automobile that in order to take us to our job thirty miles away, has to have a large enough engine to provide all the necessary power in the first ten miles so it can coast the last twenty. Obviously, the car will not coast for twenty miles, so we must add some method of storing the energy for the last twenty miles. In the case of an automobile, this could be accomplished with flywheels or batteries.

Earth-based solar systems are also going to need storage systems for the same reason. In the case of solar heating for a house, the storage system can be approached in two basically different ways. The most common concept is quite simple but suffers from low efficiency. In this method large rock beds are placed in a convenient area, usually under a house. These are heated by flowing solar-heated air through the bed during the charging cycle and extracting the heat with a forced air flow during the extraction cycle.

A much more efficient storage system employs various brines that have very high heat-storage capacity. The problem here is that the system is more complex and requires pumps, motors, controls, and piping, much of which must operate in a very corrosive environment. As a result, the maintenance cost is high. So it is hard to win. The choices are either a cheap storage system that requires large heaters or costly storage systems that can use smaller heaters.

To top off the cost story, if we do not want to have everything in our house frozen solid—including ourselves—after a five-day winter blizzard, we will have to add some type of conventional backup system. It might be gas, oil, coal, electricity, or maybe a wood stove or fireplace. In any event it adds to the cost. Even the cost of wood has increased dramatically because of increased demand.

In this example we have actually had to invest in three systems instead of one, and the primary solar system must be oversized by at least three factors. The plus side of this equation is that the sunlight is free.

The situation is not all bleak for ground-solar systems. Many areas of the country have weather conditions suitable for economical utilization of solar heating systems. The total amount of energy these systems can provide for a society such as ours, with its high technology and high standard of living, is limited because the energy is in the form of low-temperature heat. While it can be used for heating, it is not very useful for running machinery or household lights and appliances.

Solar Cells—Electricity from Light

Solar cells, or "solar batteries" as they were originally called by the scientists of the Bell Telephone Laboratories who developed them, are solid-state, photovoltaic devices that convert sunlight directly into electricity with no moving parts. The initial cells were made from pure silicon wafers with certain impurities and electrical contacts added to each side to give the cell its unique characteristics. When the cell is exposed to sunlight the light dislodges electrons from atoms in the cell material. As the negatively charged electrons flow to one side of the cell, the other side gains a positive charge from the deficiency, creating an electrical charge between the connectors. When the contacts are connected to an electrical circuit, current will flow as long as the cell is exposed to light and the electrical circuit is maintained.

The development activity undertaken during the 1970s was focused on two major areas: conversion efficiency and cost. These two categories were addressed in several different ways, which included testing alternative materials, processing refinements, multiple layers of material, thin films, single-crystal versus amorphous material, cell size, and manufacturing processes.

A great deal of progress was achieved during that time, and numerous types of cells were developed while refinement of single-crystal silicon cells continued. Additional development has occurred since that period as more and more applications have been found. Worldwide output of solar cells has increased fifty-fold since 1978.

Single-crystal silicon cells are still the most common. Typical efficiencies have been steadily increasing from about 7% several years ago to today, when 13% to 14% efficiency is readily available from many commercial outlets in various sizes of pre-assembled, sealed panels. Numerous small, low-power devices, such as pocket calculators and ventilator fans, are also run with small silicon solar cells. Development of single-crystal silicon cells continues, with maximum efficiencies of 21.6% being achieved in the laboratories.

The highest solar cell efficiency obtained to date was accomplished by Boeing researchers with two-layer gallium arsenide and gallium antinomide cells that were 32.5% efficient at converting sunlight in space to electricity. Sunlight in space has a higher energy level than on the earth, but it also contains a lower proportion of the red light spectrum (due to the increased proportion of ultraviolet light that the earth's atmosphere filters out). Therefore, it is more difficult to achieve as high an efficiency of conversion in space as on the earth. The use of multilayer cells provides for capture of a broader range of the light spectrum. Even though it is more difficult to obtain as high a conversion efficiency in space as on the earth, the *total* energy generated is much higher.

Progress in polycrystalline (multiple crystals) silicon cells has also been spectacular, with efficiencies of 16.4% in laboratory tests being reported by Japanese researchers at Sharp. Polycrystalline silicon is less expensive than single-crystal silicon. Another approach to silicon cells is being pursued by Texas Instruments and Southern California Edison. They are experimenting with silicon bead cells that can be made at very low cost, but their efficiency is only about 10%.

Another type of cells are called thin-film. These are cells that are only a few microns thick. A micron is one millionth of a meter, or approximately four one-hundred-thousandths of an inch. To put that in perspective, a human hair is about 75 microns thick. The advantages of thin solar cells include their light weight and the reduced amount of expensive materials required to make them. One type of thin-film solar cells is made from multiple layers of material, starting with an electrode, then P-type cadmium telluride, next N-type cadmium sulfide, topped with a transparent electrode and cover glass. The entire thickness is six microns—twelve times thinner than a human hair. Ting Chu, a retired professor at the University of South Florida, with the cooperation of researchers at the National Renewable Energy Laboratory, achieved a breakthrough efficiency of 14.6% with this type of cell. That threshold was soon broken, and they have now reached 15.8% efficiency with more advances likely.

Another of the thin-film cells that many believe has the most potential of all is made from copper-indium-diselenide. They have been made in meter-square sizes at 10% efficiency. The National Renewable Energy Laboratory has achieved 16.4% efficiency in the laboratory with copper-indium-diselenide and expects to increase that with future developments.

A unique new concept called roll-to-roll solar cells has been developed by Stanford Ovshinsky and his company, Energy Conversion Devices. This concept starts with a sheet of stainless steel coated with silver and three different layers of transparent thin-film amorphous-silicon cells topped with a transparent electrode. This type of solar cell can be made in continuous strips and has nearly 14% efficiency, also with a thickness much less than a human hair.

A totally different approach is to use concentrators to focus a large area of sunlight onto a small area of solar cells. This concept uses low-cost concentrators ("fresnel lens") and high-efficiency solar cells, which can have higher costs because of the reduced number required. The conversion efficiency is also increased by concentration of the sunlight. A single crystal silicon cell with 20% efficiency without concentration increases to 26% with high concentration of sunlight.

As development has progressed, efficiency has increased and cost has decreased. One of the major problems with solar cells in the 1970s was the lack of large-scale, low-cost production techniques, but as the years have passed there have been major developments to improve mass production techniques. Today some of the manufacturers have highly automated assembly lines and are rapidly lowering the cost of their delivered solar cell modules. The lowered cost and higher efficiency has greatly enhanced the attractiveness of solar cells as an energy generating source. Unfortunately, no matter how low the cost, terrestrial-based solar cells cannot overcome the same problem as other earth-based solar systems—intermittent sunlight.

Solar cells have powered our space satellites for many years. In the space environment they are exposed to the full energy of the sun; however, when sunlight passes through the atmosphere, even directly overhead, it loses more than 25% of its energy to the atmosphere. This loss is greatly increased as the sun nears the horizon. As a result, a solar cell's output varies from dawn to dusk at about the same ratio as we get sunburned on a dawn-to-dusk fishing trip. The overall result is that for the same cell efficiency, it takes about five times as many solar cells on the ground as on one of our communications satellites to generate the same amount of power—and that is if you lived in the Mojave Desert. If you lived in "Average Town, America," it would take 15 times as many cells; if you lived in the Pacific Northwest, as I do, it would take 22 times.

In addition, in order to be utilized effectively for 24-hour-a-day energy, we need to provide a storage system and a power processor to correct voltage variations or electricity form. Solar cells generate direct current (DC) electricity, and our homes currently operate on alternating current (AC) electricity. If we lived in "Average Town, America," and have an average household that uses electricity for cooking, lighting, and household appliances but not for either heating or hot water, and if we wanted to power our residence with solar cells, it would be necessary to cover an area equal to the area of our roof. That area would also have to rotate to track the sun from dawn to dusk, or else be even larger. In addition, we would have a bank of batteries in our garage and a power processor humming away in the corner. We would be suffering from being on the wrong end of the effects of scale. We would have created in a small area a complete power plant that could perform all the basic functions of a large, community plant. We would be paying all the costs ourselves instead of sharing many of the costs of common functions with other users, as is the case of a utility selling to many customers from a common facility.

One approach being used by the Sacramento Municipal Utility District to minimize some of the problems is to mount four-kilowatt solar cell arrays on the roofs of private homes, but instead of feeding the power directly to the houses, it is fed directly into the power grid. In this way it is used to contribute energy to the entire power grid during the day to help supply the daytime peak loads. The homes themselves obtain their power from the normal grids—and the residents pay a premium for the privilege of contributing power to the grid!

The future for distributed generation of electric power is not all bleak. Many remote locations find solar-cell-generated electricity cheaper than any other approach. Sunlight is free, so there will be many places where the cost of bringing centralized power to a remote site is greater than the added cost of distributed systems. On many remote islands of the South Pacific the only source of electrical energy is solar cell panels or an occasional Honda gasoline generator.

Centralized Solar Power Generation

One concept for centralized power generation from ground-based solar utilizes fields of mirrors (or heliostats, as they are called) to concentrate sunlight into a cavity absorber or receiver. When the sunlight is concentrated in the cavity, the temperature becomes very high and heats a fluid that is pumped through a heat exchanger inside the cavity and then routed to a turbine engine. The concept is essentially the same as a coal or gas-fired power plant, except that the heat is provided by sunlight instead of by fire. There are also variations in how the working fluid is heated. One other approach uses trough-shaped mirrors concentrating the sunlight on a pipe carrying the heated working fluid. Since they depend on heat from the sun, they suffer from many of the disadvantages of other ground-solar applications, but with three major differences.

First, they can be built in the best locations for sunlight. Second, they can provide electric power to a primary distribution grid that has a higher demand during the day than at night. Third, they can take advantage of the cost reductions associated with a large-scale system. By using thermal engines, they convert energy more efficiently than solar cells. However, in regard to absolute power costs, they cannot avoid the fact that the sun sets at night and the plant lies idle until the morning. There is an option of burning a fuel, such as natural gas, at night to run the generators, which is done at most locations to help balance the profit-and-loss ledger. The major existing example of a ground-based solar power plant is at Daggett, California, where the facility generates about 400 megawatts of power.

The use of solar cells is also being applied to central power stations that use large fields of solar cell panels to generate electrical power. Currently power costs of about 25 cents per kilowatt-hour are achievable with a goal of 6 to 12 cents per kilowatt-hour by the year 2000. This would make solar cell electricity competitive with existing systems for the periods the sun is shining. This will make centralized earth-based power plants quite attractive for the areas where peak demand is during the day while the sun shines.

Windmills—The 24-Hour Solar System

Windmills, or wind turbines—the modern resurrection of wind as a source of energy—have an interesting aspect that we do not often consider. Wind is one of nature's storage devices for solar energy. It is a product of the earth's rotation and air currents generated by the heat of the sun. Sunlight heats the air and the earth's surface each day, causing convective currents to be generated. Since the atmosphere is a global phenomenon with worldwide interactions, and the sun is always shining on half the globe, we experience wind in some form day or night. The only problem is that nature is fickle and lets the wind flit around from place to place. Even so, there are some locations that have fairly consistent winds.

Modern airfoil (propeller) technology has been used, and wind turbines of three-megawatt peak capacity have been built, but the dynamics of such large rotating propellers cause long-term fatigue problems beyond the capability of the materials being used. Smaller units do not experience this degree of difficulty and have been built and installed in large numbers in a few selected locations. They are making a measurable contribution to the power grids. As an example, there are 7,000 wind turbines installed on the slopes of California's Altamont Pass. Initially they were able to compete because of governmental incentives, but with time and numbers of production units they have reached the point where they can nearly compete with other sources without the incentives. They can only nibble at the problem, however. They simply cannot be made big enough or placed in sufficient numbers to completely swallow it.

Biomass—Liquefying Nature's Solar Fuels

Another odd solar power energy source is biomass. It can be considered in many forms. It may be the controlled digestion of garbage to produce methane gas and other products; it could be the distillation of grains or sugar beets to alcohol for gasohol; the growing of special crops to be used in biomass conversion to fuels; processing of livestock manure; the collection of wood scraps; or the planting of fast-growing trees for chemical processing into fuels. These are all examples of biomass conversion.

Biomass conversion is actually the last liquid-fuel-producing step after nature's solar energy process grows the fuel. This process starts with the growing of plants of various types, all of which grow because of nature's photosynthesis process—using solar energy. Unfortunately, nature's process represents only about 1% solar efficiency—not nearly high enough to sustain massive demands from an industrialized society. This has already been demonstrated in the past when wood was the primary fuel used in England. Biomass conversion to further process wood or its equivalent to liquid fuels will add more inefficiencies. The net result is that these fuels have limited capacity and can be cost-competitive only in very select markets. An example is Brazil where there is extensive use of gasohol. They have sufficient feed stock available to make gasohol cost effective. It also is energy generated within the country, which helps their foreign debt problem.

One exception to the low solar efficiency of nature's conversion cycle has been found in purple pond scum. The purple bacteria that float in stagnant water convert sunlight to energy with 95% efficiency, more than four times that of solar cells. Maybe we can figure out how to convert purple pond scum into something useful or at least find out how to duplicate the process. While it is only a glimmer in some scientist's eye at this point, who knows what might happen in the future.

Biomass fuel can be a lucrative by-product of processes that are necessary for other reasons. Methane gas and other fuel gases are being collected from old landfills and are added to industrial and community gas supplies. There will undoubtedly be many instances where this type of development can be economically attractive and beneficial to the general welfare and the environment. The total contribution of this energy source is limited, however, to nature's rate of renewal.

Ocean Thermal Gradient—The Hard Way to Run an Engine

The last earth-based solar system considered during the 1970s was ocean thermal gradient, which uses the temperature differences found in the oceans. Whenever a steady-state temperature differential exists between two sources, there is a possibility of operating a thermal engine. The greater the two temperature extremes, the easier this is to accomplish and the more mechanical energy can be extracted. The situation really becomes challenging when the two temperatures are close together, as would be the case with ocean thermal gradient power generation.

The concept is to use the warm solar-heated water near the surface of the ocean as a heat source and the cool water several thousand feet down as a heat sink to reject heat. The problem is that the maximum differential is only about 45 degrees Fahrenheit. As a result, massive amounts of water must be moved to provide significant amounts of power. Moving massive amounts of water from the surface to a depth of 3000 feet, or vice versa, takes huge equipment. All of this equipment would be moored in deep water at selected ocean sites that had sufficient temperature differentials.

Maximum efficiencies believed to be achievable would be about 3%, with realistic efficiencies probably closer to 1%. The question of economic viability pivots on whether it is possible to build and maintain such massive facilities in the hostile environment of the sea for sufficiently low cost—it is not likely.

Geothermal—Tapping the Earth's Energy

A natural source of heat is the very heart of our earth. Only the surface is cool, and as we penetrate deep underground the temperature rises. At some natural locations, the high temperatures come very close to the surface. We are aware that the eruption of Old Faithful Geyser in Yellowstone Park is caused by the heating of water to steam, building up pressure that forces the geyser to erupt at regular intervals. In fact, Yellowstone Park is covered with many reactions caused by geothermal activity close to the surface.

In some areas of the country, similar types of conditions occur and some have been developed as energy sources. There is a 600-megawatt plant at The Geysers, California. This site was easy to develop where natural activity is close to the surface. New Zealand generates 6% of its electric power from geothermal sources rising close to the surface. However, they are experiencing a gradual reduction in output as the area is cooling. This is also happening at The Geysers in California. The real question is whether we can tap the earth's core heat on a large scale.

Concepts have been developed and some test work accomplished, but the case for large-scale development of geothermal does not look good. Engineering reality must be considered, which means that to extract large amounts of energy it is necessary to pump large amounts of water down to the high temperature areas. While it is down deep in the earth it must be heated. It takes a large area to accomplish the heating of a large volume of water. While this heating takes place, the surrounding area is being cooled by the water. Since rock is not a very good heat-transfer medium, we would soon find that we have cooled off our heat source—such as is happening in the New Zealand plant—and we would have to drill new holes.

Development work continues, however, and some small power plant sites are being developed. Large-scale development is not likely unless new methods are developed to economically tap the heat that lies beneath our feet.

Fusion Power—The Elusive Carrot

The promise of fusion power is the carrot that has been dangled in front of us since research for power generation was started in 1951. Fusion is the combining of atoms, the opposite of fission, which is the splitting of atoms. Fusion is the energy of our world, of our life. Our very existence depends on it. The sun and the stars are all operating fusion reactors. Without the sun we would not be.

We have accomplished fusion reactions on the earth in the form of hydrogen bombs. This is achieved by using a fission reaction to raise the temperature and pressure high enough to cause the combining of the hydrogen atoms in a fusion reaction to form helium. This is not a very "user friendly" form of energy if you have to set off an atomic bomb in order to make your morning toast.

Research has been going on throughout the world since 1951 to develop a method of achieving a controlled fusion reaction that can be harnessed into generating electrical power. The great advantage would be nuclear power using hydrogen as a fuel without the radiation dangers of fission reactors. The difficulties in achieving a controlled reaction are immense. Many breakthroughs have been announced, but the researchers have yet to achieve a reaction that could be used to extract useful energy.

For a while, March 23, 1989, was thought to be a date that would go down in history as one of the great milestones of our time. It was on that day when B. Stanley Pons, chairman of the University of Utah's chemistry department, and Professor Martin Fleischmann of Southhampton University in England announced that they had carried out nuclear fusion through an inexpensive and relatively simple electrochemical process. Pons said the technique could be used to supply nuclear energy for industrial and commercial use. The announcement that they had achieved "cold" fusion in a laboratory jar generated tremendous excitement (and skepticism) among the scientific community. If true, it would be a dramatic breakthrough in the four-decade search for a method of controlling sustained nuclear fusion.

Alas, it was not to be. Later testing by other researchers failed to achieve the same results and revealed the errors made in the experiments. The hopes raised by Pons and Fleischman fizzled along with their scientific careers.

Fusion power research goes on at high levels of expenditure, and the predictions continue that it is only 20 or 30 or maybe as long as 50 years in the future, but it seems that the goal is more elusive than ever. The carrot is becoming limp and stale indeed.

Space-Based Solar Power—The Dark Horse

In 1968 when the United States was deeply immersed in the final testing of the Saturn/Apollo space vehicles for sending men to the moon, one farsighted visionary conceived an idea of a way to use space as a place to gather energy for use on the earth. Dr. Peter E. Glaser of Arthur D. Little Company first proposed the concept of placing satellites in geosynchronous orbit to provide energy from the sun. He saw them covered with solar cells to generate electricity, provided with an antenna to transmit the energy to the earth in the form of radio frequency energy waves, and then reconverting the energy back to electricity. This was based on the concept of wireless energy transmission first demonstrated by William C. Brown when he successfully powered a model helicopter with a wireless radio frequency energy beam.

The idea sounded like the invention of a science fiction writer at the time, but it was based on sound engineering principles. A few space-engineering enthusiasts around the country started looking at the concept in more detail, and by the early 1970s, several small studies were being conducted by aerospace companies and NASA. After the 1973-74 oil embargo these studies were expanded and culminated in the DOE/NASA systems definition studies of 1977 to 1980 that I described earlier.

The outcome of these studies followed Dr. Glaser's original concept but provided greatly expanded design definition, understanding of the technology, and in many areas, test data. Each step of the required technology is in use in some form, for other purposes, somewhere in the world today. There are no scientific breakthroughs required. Even so, the engineering task of designing and developing such a concept is immense. However, the long-term potential benefits are even more immense.

What is the Solution?

I have reviewed briefly what energy concepts have been studied or developed in the search for new energy systems to replace or at least contribute to the replacement of oil as our major energy source. None of them have as yet emerged as the energy source for the next energy era. The question is whether any of the known concepts can give birth to the fourth energy era.

The test is to measure each system against the criteria.

Conservation—A Dead-End Street

Conservation is clearly the quickest and easiest way to eliminate the use of oil. Where can we buy another gallon of gasoline or another kilowatt of electricity for the cost of simply using neither? The price is zero dollars and there is no pollution. Unfortunately, cost and pollution are only part of the problem. Conservation cannot power industry, grow food for the emerging masses, light and heat our homes, cook our food, nor power the transportation system to bring us to our jobs and vacations. To accept conservation as the solution is to simply give up on the future of humanity and our planet. It is a stop-gap measure to buy us time, but it cannot meet any of the criteria for the world's future energy needs. We would revert to subsistence levels of existence after the deaths of billions of people from starvation and war as scarcity forced country after country to desperately reach out for the essential elements of life.

Many people believe strongly that conservation is the only solution. They feel guilty about what we have here in America and think the "right" way is not to use energy so that they can share the burdens of so many poverty-stricken people in the rest of the world. Unfortunately, this does not help the poverty-stricken people, but rather makes their future prospects even grimmer and ensures the fact that they can never improve their lives.

This does not mean that increasing energy efficiency is bad. However, when the goals go beyond increasing efficiency they become counter-productive and destructive. Most of the gross inefficiencies have now been eliminated. It is time to get on with solving the problem of how we are going to provide the clean low-cost energy needed to power our society and help the other people of the earth to climb out of their pit of poverty.

Coal, Oil, Natural Gas, and Synthetic Fuels—A Fading Future

Any of the systems based on fossil fuels such as coal, oil, natural gas, and synthetic fuels share the problem of being finite resources and are subject to ever-increasing cost as they grow ever more scarce. Even though there are certainly some reasonably large reserves left in the world, they are ultimately limited. It does not take much more of a demand than there is supply to cause large price increases. Increasing world population and requirements from emerging underdeveloped countries will also increase overall demand.

The other key issue is atmospheric pollution. The United States has established controls to help minimize the amount of pollutants, but even with tight controls it is a very serious problem. Many countries have not made any attempt to control pollutants. In any event, scrubbers cannot remove carbon dioxide from the smokestacks, and the level will continue to build in our atmosphere until we stop the excessive burning of fossil fuels. The fossil fuels fail to meet the criteria of nondepletability, environmental cleanliness, and low-cost over the long term. Fossil fuels must be replaced for many of our energy needs so that they can provide the fuels for the systems that cannot readily be powered with electricity.

Earth-Based Solar—The Struggling Contender

Earth-based solar systems can meet four of the five criteria—the source is nondepletable, environmentally clean, available to everyone (wherever the sun is shining, that is), and convertible to a useable form. There is some possible question about the effect of land use if earth-based solar systems had to supply large amounts of power, but their biggest problem is cost. Large-scale development could help, but the fundamental stumbling block is the intermittent supply of sunlight, which creates the need for very large-scale facilities and also the requirement for some type of energy storage devices. To put the problem in perspective, I offer a very simple example. If we were to use solar cells that were 15% efficient at high noon at the most optimal earth location, we would find that the actual overall conversion efficiency of sunlight to usable electric power, averaged over a year, would be less than 3%. But if we were to locate the solar cells in an average US location, they would only be 1% efficient overall. That does not even take into account the inefficiencies of the storage system that would be required for when the sun wasn't shining. When conversion efficiencies drop below 3%, the cost of the equipment becomes a dominant factor. Higher conversion efficiencies are certainly possible and even quite practical, but the cost of the equipment is also increased and the systems must be maintained in the unprotected outdoors.

The use of ground-based solar systems is very likely to expand, particularly for remote areas, but unless there are cost breakthroughs that are unforeseen now, the price will be too high for massive use. If there is no other solution, earth-based solar systems can be expanded to supply world requirements, but the cost of the energy will limit world economic growth. It is a contender, but it is struggling against heavy odds.

Nuclear—Had its Chance

Nuclear fission is an energy source that would come close to meeting the requirements if certain conditions could be met. The first is to eliminate emotion, fear, and prejudice from the decision-making process. Second, we would have to develop and use a breeder reactor. Third, we would need to develop an effective plan for waste disposal. None of these are likely to happen. And unless the breeder reactor is used, there is not enough uranium on the earth to meet the criteria of nondepletability.

When it comes to cost, the issue of nuclear power is clouded. When the atomic age began back in 1945, proponents of nuclear power talked of energy so cheap that it would not be necessary to meter it. We all know that the reality has been quite different. We should not dismiss this issue too quickly, however, for it exposes a fundamental problem. This is the difficulty of converting energy in one form, which is not useful, into a form which is useful. If we could use the total energy available from splitting the atom, the statements made decades ago would be true. We cannot use the energy directly, however, unless we want to make a bomb. Therefore, the cost that we now experience is heavily driven by the cost of the machinery and facilities to convert the heat of a controlled reaction first to mechanical energy and then to electricity. This drives the cost higher because of the gross inefficiencies in the thermal process—inefficiencies driven by material limitations and technological limitations. We are only able to convert an extremely small fraction of the energy released by splitting the atom into useful electricity—about 1%. As is the case with earth-based solar systems, it is the conversion process that causes most of the cost.

The costs of breeder reactors are projected to be about 30% higher than conventional fission reactors, and they require fuel reprocessing. As a result, the future energy costs are questionable.

The environmental issue regarding breeder reactors is as emotional as it is factual. The emotions react to the safety questions of nuclear proliferation, radiation danger, and waste disposal. All of these are serious issues that have not been satisfactorily answered. That will not be easy in light of the accidents that have occurred and the evidence of gross negligence in handling nuclear materials becoming evident in the former Soviet Union.

Strangely enough, though, one of the truly significant environmental issues is seldom ever raised. This is thermal pollution. Whenever thermal energy is converted to mechanical energy—as is the case in nuclear reactors or coal and oil plants—two thirds of the thermal energy is waste heat and must be disposed of. This can become a serious problem. Today's solution is to dump it into the atmosphere and the waterways. Have you seen the great pillars of steam rising from the cooling tower of an atomic plant or coal plant? That is thermal heating in action.

So we find huge question marks when we try to measure nuclear fission against our criteria for a new energy source. It is not a solution that has the support of the people, and it is not likely to improve its position with time. It has had its chance and failed.

Nuclear Fusion—The Limp Carrot

The most likely place to find a potential solution to our energy crisis is with an advanced technology approach. Fusion is certainly an advanced technology. Unfortunately, it has not advanced to the point that a realistic fusion plant can be defined. In theory, there should not be any of the radiation problems associated with fission reactions, and if hydrogen can be used as a fuel, then it can be considered nondepletable for all practical purposes (since we can always separate hydrogen from water). Theory is often very difficult to put into practice, however, and in the case of fusion, this is doubly true. The pressures and temperatures required to start a fusion reaction are far beyond normal human experience and the materials we possess, so whole new sciences must be developed and applied. Besides, research has been directed primarily toward achieving a controlled reaction, not to power plant design.

Because of the unknown scientific and engineering features in the fusion power concept, it has been impossible to adequately characterize an operational power plant. As a result, it is not possible to make meaningful estimates of facility and operational costs. Without knowing these costs, the cost of fusion power cannot even be estimated.

Fusion power could be the clear leader of the fourth energy era *if* there are some scientific breakthroughs and *if* the energy can be converted directly into electricity without going through a thermal engine conversion. Anything can happen, but it has not happened yet.

Ironically, the Department of Energy has been focusing on developing fusion reactors to generate power that still has to be converted to electricity to be usable. In the meantime, we've already got a fully operational fusion reactor a safe distance away—93 million miles—that should continue to function for another four billion years. All we have to do is find a way to convert that fusion power to electricity—a job for which solar power satellites is eminently suitable.

Solar Power Satellites—Hope for the Future

The last of the candidates—solar power satellites—is the one that can best meet all of the criteria to become the energy system of the next energy era. I will review how it satisfies each criterion.

Low Cost

The first criterion is low cost over the long term. Usually the first reaction to the question of a solar power satellite being low cost is that nothing associated with space could possibly be low cost. That is simultaneously a correct reaction and an erroneous one. It is correct when considering the cost of hardware designed to operate in space on an independent basis with high reliability. Based on dollars per pound of space hardware versus dollars per pound of, say, a spool of copper wire, there is no comparison. But that very same piece of space hardware—a communications satellite, for example—can reduce the cost of an international telephone call by a factor of ten times less than could be provided by the spool of copper wire strung from one point on earth to another point far away. Now which one is lowest cost?

The same principle applies in the case of the solar power satellite. The hardware is not cheap, but it has high productivity. The high productivity is achieved because the solar power satellite is in the sunlight over 99% of the time, which is five times more sunlight than is available at the best location on earth. It can operate at maximum capacity at all times and does not need a storage system. Its overall efficiency of converting sunlight to electricity delivered on earth is projected to be from 7% and 10%, and the system will be operating in the benign environment of space. This compares to the 1% to 3% for earth-based solar cell systems, and along with the favorable environment and the elimination of storage systems, is the fundamental reason for going to space for solar energy.

If we use the cost estimates established from the preliminary designs developed by the NASA study contractors in the late 1970s, then the cost of power would be less than the cost of electricity generated by coal, oil, or nuclear power. When the initial capital costs are paid off, the cost of power could then drop to a fraction of the costs from other sources. The power costs are at least in the right ballpark. The energy is free; the only cost is the cost of the conversion hardware and the cost to maintain it. The environment in space is very favorable for most equipment. There is no wind or rain or dirt or oxygen or corrosive fluids. Things last a very long time in space. The potential exists for long-term low cost—without the inflationary cost of fossil or synthetic fuels.

Nondepletable

Second is the question of depletability. It is clear that the energy source is nondepletable since it is available for as long as the sun shines and, therefore, for as long as man exists. Only one part in two billionths of the sun's energy is actually intercepted by the earth. This extremely small fraction is still a massive amount of energy. The satellites would not even have to infringe on this increment, however, as they would intercept the energy that normally streams past the earth into deep space. Geosynchronous orbit is about 165,000 miles in circumference—ample room to place as many satellites as we desire. The amount of energy that can be gathered and delivered to earth is primarily a function of how much we want, and only the usable energy is delivered to the earth.

Environmentally Clean

The environmental issue is what has stopped the construction of more nuclear power plants. Can solar power satellites pass this criteria? First of all, it is difficult to fault the energy source as environmentally unacceptable, even though most dermatologists try. The rest of us think having the sun around is just fine. Putting the power plant and its associated equipment 22,300 miles from the nearest house does not seem like a bad idea, either, especially when the thermal loss of energy conversion is left in deep space and will not heat up our rivers and atmosphere as all the thermal plants do.

But what about the wireless energy beam? Is it a death ray that will cook us if something goes wrong and it wanders from the receiving antenna? No. Even though the radio frequency beam is the same kind of frequency as we use for cooking in our microwave ovens, the energy *density* (or the amount of energy in a given area) is much less than the energy density in our microwave ovens (because our ovens are designed to contain the energy and concentrate it within the oven cavity). In fact, the wireless energy beam's maximum energy density would be less than ten times the allowed *leakage* from the door of a microwave oven. At that level, which would be a maximum of 50 milliwatts per square centimeter, a person would just feel some warmth if he or she was standing in the center of the beam on top of the rectenna (not a very likely event). That much energy is less than half of the energy found in bright sunlight at high noon on a Florida beach, except that it is in the form of high-frequency radio waves, or microwaves. The only definitely known reaction of living tissue to microwaves is heating.

There is much debate about other possible effects, such as nervous system disorders or genetic effects due to long-term exposures at low levels. No good, hard evidence exists to prove or disprove the allegations. Many studies have been made and others are underway, however, to try to clarify the issue. In the meantime, let us consider the general evidence accumulated over the last century.

X-rays and the natural radiation of radium were discovered at about the same time as radio waves. In fact, Wilhelm Rontgen discovered x-rays in 1895, which was the same year that Marconi invented the radio telegraph. As early as 1888 both Heinrich Hertz and Oliver Lodge had independently identified radio waves as belonging to the same family as light waves. The big difference between nuclear radiation and radio and light waves is that radio and light waves are non-ionizing, whereas nuclear radiation is ionizing. Unfortunately, people often confuse the two. During the ensuing years, it became very clear that the magic of x-rays and the natural radiation of radium went beyond what was originally thought. Serious side effects were soon discovered. Mysterious deaths occurred among workers who painted the luminous dials of watches. The development of the atomic bomb lead to the discovery of many more effects of excessive exposure to ionizing radiation.

During that same period, radio, radar, and television grew at an even more rapid rate. Radar, television, radio, and space communication frequencies spanned the entire radio frequency range. Energy systems were added among the communication frequencies. During all these years of exposure by everyone on earth, the only nontransient effect identified has been heating. The point I am making is that if some serious phenomenon were caused by radio waves, there should be indications by now.

The overall picture for the microwave environmental issue looks good, but additional data will be needed to be certain. This is the hardest data to gather—information to prove that there are no effects.

The companion environmental issue is the question of the land required for the receiving antenna. Because the energy density is restricted to a very low level in the beam—in order to assure safety—the antenna must be large in order to supply the billion watts of power from a solar power satellite. The antenna would be about 1.8 miles wide. Since it can be elevated above the ground and since it would block less that 20% of the sunlight while stopping over 99% of the microwaves, the land can be used for agriculture as well as for the receiving antenna. In comparison, the total land required is less than with most other energy systems. The amount of land required for the receiving antenna is actually much less than that required for coal strip mines to produce an equivalent amount of power over 40 years.

Available to Everyone

The satellites may be located at any location around the earth and would be able to beam their energy to any selected receiver site except near the North and South Poles. Certainly they could make electricity available to all the larger populated areas of the earth, if those areas purchased a satellite or bought the electricity from a utility company that owned one.

It is not possible for most countries to be able to afford the development costs of a satellite system, but once developed the cost of individual satellites would be within the capability of many countries.

In a Useful Form

With solar power satellites, the form of the energy delivered is electricity, the cleanest and highest form available to us. It is the form we need to clean up the earth's environment. It is the energy form of the future.

Here at last is a nondepletable, clean energy source with vast capacity, within our capability to develop, waiting to carry us into the twenty-first century.

7

Electricity: The Energy Form of the Future

Since the beginning of man's experience on this earth, he has stood in wonder and fear as lightning laced the sky and thunder drove the terror of the unknown into his soul. What was this fearsome force that could make night like day, strike great trees asunder, burn forests, and occasionally strike a victim dead in its path? The ancients attributed this frightening phenomena to the wrath of their gods. It was not until the eighteenth century when Benjamin Franklin conducted his famous experiments that fact started to find its way into the mystery of lightning.

For most people at that time electricity still carried that strange mysticism. What good could it be? What could it be used for? It was a scientific curiosity for many years, but after the start of the nineteenth century our comprehension of its potential began to grow rapidly.

The real understanding of electricity started in 1800 when Alessandro Volta produced the first electricity from a cell made of zinc and copper plate. This was the first battery, but it was much more than that. Of greater significance was the accompanying discovery of electrical current, which began the development of modern electrical science and industry. In Volta's honor the term "volt" was given to electrical pressure, or electromotive force. In the early days of the century the only source of direct current was from primary batteries of the type Volta had invented, and they were used extensively.

Later, in 1819, Danish physicist Hans C. Oersted discovered electromagnetism. In 1820, Andre Ampere defined the laws of electrodynamic action, and in 1821 Faraday discovered the fundamentals of electromagnetic rotation. In 1827, George S. Ohm made his contribution when he formulated Ohm's law, which defined electrical current potential and resistance. This was all put together in 1829 by American physicist John Henry when he constructed an early version of the electromagnetic motor. This was the beginning of making electricity perform useful work. The first effort to propel railroad vehicles by electrical batteries was made in 1835, but it was not until 1879 that E. W. Siemans exhibited the first successful electric tram at the Berlin Trade Exhibition.

Electricity branched off into the communications world in 1833 when K.F. Gauss and Wilhelm E. Weber devised an electromagnetic telegraph that functioned over a distance of 9,000 feet. Wheatstone and W.F. Cooke patented an electric telegraph in 1837, but it was Samuel Morse who exhibited his electric telegraph at the College of the City of New York in the same year, receiving a $30,000 grant from Congress to build the first telegraph line from Washington DC to Baltimore. It was used for the first time in 1844. The electric telephone was brought into existence by Alexander Graham Bell in 1876, and the world suddenly became much smaller.

By the time Thomas Edison and T.A. Swan independently devised the first practical light bulbs in 1880, electricity's future was secured. Edison was also the man who did the design of the first hydroelectric power plant, which was built at Appleton, Wisconsin, in 1882. The first English electric power station was established at Deptford in 1890. Niagara Falls became more than just a honeymoon site when hydroelectric installations were begun in 1886 and delivered their first power in 1896.

The scene was set for the twentieth century to become the century of electricity. New uses were discovered steadily, and there seemed to be no end to what electricity could do. Communication systems depended on it, and transportation became involved with electric trolleys, buses, trains, subways, and a few cars. The wheels of industry started to turn with the torque of electric motors. Aluminum was changed from a metal once so precious that Napoleon had his tableware made from it, to a metal so common we now use it for beer cans, pots and pans, window frames, and airplanes.

The first all-electric city, a place with no chimneys, was built as a model at Grand Coulee Dam in the mid 1930s. It demonstrated the flexibility of electricity to perform all household energy chores more than half a century ago. Since that time, technology has made great strides, and today we take for granted the many tasks it can accomplish at very high efficiencies.

The Flexible Energy Form

Through the years, the portion of our total energy use supplied by electricity has steadily grown from about 12% in 1945 to more than 30% in 1980. This trend is on an accelerating rate for two fundamental reasons. First is the great flexibility of electricity to do nearly any energy job due to the high grade of electrical energy. Second, as the cost of oil rises, the cost of electricity has risen more slowly because only a small fraction of our electric power is now generated with oil. The rest comes from coal, natural gas, nuclear, and hydroelectric dams.

You are probably asking yourself what the term "high-grade" energy means. It might be easiest to start by defining low-grade energy. Low-grade energy is typified by heat at relatively low temperatures such as that found in the heating elements of our hot water tanks. Low-grade energy is useful to heat water for our homes, but not very useful for anything else.

As we move up the scale of energy and consider high-temperature steam as an example, we find that it can do many things. It can run engines and provide process heat, among other things. When we look at electricity, we find it is the highest grade energy of all because it can do nearly anything. Electricity can provide thousands of degrees in the arc of a welding torch, activate the sensitive heating elements that defrost our car's rear windows, run the quiet motor of the refrigerator-freezer in the kitchen, control the ever-watchful thermostat maintaining our comfort, bring us lilting stereo music in the living room, energize the comforting whirl of our car's starter as it brings the engine to life, light our Christmas tree, bring us the sound of our children's voices from far-off places, show us the thrill of an Olympic winner from halfway around the world, light and heat our homes, and run the power tools that make life so much easier. In the United States more than 98% of all physical work is done by machines. Most are powered by electricity. Can you think of any other energy form that can perform the variety of functions within the capability of electricity? There simply isn't any.

The Clean Heat

Now that you are an expert on the grades of energy, it might occur to you to ask: "Why does it make sense to use a high-grade energy like electricity to do a low-grade energy job like heating my home?" That is a very sensible question and has more than one answer. First, if you have low-grade heat available from solar energy or leftover heat from the local foundry, it only makes sense to use it. However, the great advantage of high-grade energy is its ability to appear in many guises. Most of us think of glowing coils of wire hanging on a ceramic housing in our portable electric heater when we think of electric heat. That is certainly where electric heaters started, but much has happened over the years. Using the same principle of electrical resistance in a wire, the original electric heaters have emerged as baseboard heaters, forced-air furnaces, or hot water furnaces. They have provided clean, quiet, labor-free heat to millions of people. They do suffer inefficiencies in the process, which has stimulated engineers to find a new way to turn high-grade energy into heat at very high efficiencies.

Their efforts have produced a very efficient device called the heat pump. It is just an air conditioner running backwards—not the machinery, but the airflow. The concept is very simple. It takes cool air from outside, raises its temperature by passing it through a heat exchanger that is heated by compressing a working fluid in the heat pump with mechanical energy, extracts some of the residual heat to be used inside, and then exhausts the cooled air back to the outside. In this mode, it is a heater. If you reverse the process, it is an air conditioner. The system can be made to work at very high efficiencies. There is no wasted heat going up the smoke stack, there are no combustion products to foul our environment. Heating or cooling is there with the flick of a switch. No need to call an oil truck to fill the tank. No sweaty, dirty job of shoveling coal into the coal bin as I did when I was a child. No worry about the pilot light going out on a gas furnace. With the simple act of setting a thermostat we enjoy a clean, comfortable environment.

Electric Transportation

Near the turn of the century when horses were being replaced with the infernal machines of man's creation, there was no clear consensus of what kind of engine would power these machines. Some were powered with internal combustion engines using gasoline as fuel. Some had steam engines. Some were electric with battery power. The competition raged for several years, with advocates of each touting their advantages. In the end, the winner took all. The flexibility, range, performance, and low cost of the gasoline-powered automobile could not be matched by its competitors.

This great dominance of the personal transportation market reached its peak during the 1960s and early 1970s. Gasoline was so cheap it did not really matter how much was burned. Cars were big, heavy, and luxurious with powerful engines. It was the era of muscle cars with 442- and 454-cubic-inch displacement engines that could develop more than 350 horsepower. My 1968 Firebird 400 did not even begin to run well until it reached 75 miles per hour. What auto company would even consider spending money to develop an electric car in that kind of environment? Or for that matter, what US car manufacturer would even seriously approach the development of a high-efficiency, lightweight, high-mile-per-gallon, gasoline-powered car? Certainly there would be very few buyers for them at that time.

Since then, fuel cost and availability, environmental controls, and inflation have changed all of that. As we look into the future and the new energy era based on nondepletable electricity generation, it is time to reevaluate the capability of the electric car. Many years have passed and significant progress has been made.

Battery technology has come a long way since Volta's first cell of zinc and copper plate in 1800. That first cell was a crude example of what is now called a primary cell. It was like the dry-cell flashlight batteries of today in that it could not be recharged. The secondary or storage cell had its real beginning in 1859 when the French physicist Gaston Plante is believed to have made the first practical "acid" storage battery. Storage batteries are a way of storing electrical energy as chemical energy and then delivering the energy later as electricity. They can then be charged again and again, cycling between electrical energy and chemical energy, back and forth and back and forth.

Thomas Edison invented the nickel-iron-alkaline storage battery in 1908, and through the ensuing years the two main types that evolved were the lead-acid and the nickel-iron-alkaline cells. They both operate on the same general principle but differ in their materials and characteristics. The lead-acid battery has become the one with which we are most familiar since it is used in our automobiles. In recent years we have also seen increasing use of nickel-cadmium-alkaline storage batteries, which were used extensively in Europe before their introduction to the United States.

Today batteries play an important part in our lives. Dry cell primary batteries power our flashlights, toys, cameras, personal stereos, and a multitude of other devices. Rechargeable storage batteries provide portable energy for a vast spectrum of electronic marvels, such as cellular telephones, laptop computers, and battery powered tools, as well as starting our cars and powering our golf carts. The potential of battery-powered automobiles in the future has brought new emphasis to advanced developments.

Lead-acid batteries are probably reaching the practical limits of their development since they have been in continuous use for so many years in the automobile and marine industries, as well as being used extensively in deep-cycle applications for forklifts and golf carts. Even so, they are the yardstick by which we can measure other developments, and lead-acid cells can be used effectively in battery-powered cars as long as the range requirements are modest, probably about 50 to 100 miles. General Motors is testing a new lead-acid battery-powered car called Impact that has performance comparable to many small sport coupes. It has a 70- to 90-mile range with current batteries. The designers think that new battery developments will double that range by 1998.

Edison's nickel-iron batteries are also being developed for use in a new, limited-production, electrically powered Chrysler minivan that will have a 100-mile range and a top speed of 65 mph. The expected life of the batteries, using overnight charging, would be 100,000 miles.

Another battery development that is favored by many for electric cars of the future is the sodium-sulfur technology battery. These batteries have twice the power density of lead-acid batteries. A new world distance record was set by a Swiss-built two-seat car, which traveled 340 miles at an average speed of 74.4 mph. Even though it had been specifically built for this test, it was a very impressive demonstration of the future potential of electric cars with sodium-sulfur batteries. Several major manufacturers now are using them in their prototypes, such as the Volkswagen Chico, Ford Ghia Connecta, and BMW E2.

Research is continuing on nickel-cadmium batteries. Other material combinations are also being investigated along with techniques to further improve the known material combinations. Life, weight, cost, charging rate, and safety are all important parameters to be considered.

Let us put ourselves into the picture as Mr. and Mrs. Average American and see what kind of electric car we would need in the future if we wanted to maintain at least our current standard of driving. What kind of driving do we normally do and how far do we drive?

We drive to work and home again. We go to the grocery store and to the shopping mall. We go to a ball game occasionally and stop by a friend's house whenever possible. The kids use the car to go to school functions and probably do a little street cruising afterwards. We drive to the other side of the state to see relatives at least a couple of times a year. We drive to Disney World in Florida for our vacation. We also like to go to the movies regularly and maybe go fishing in the summer.

When we add it all up, we find that of the total miles most people drive in a year, 50% is on trips of less than 20 miles and 70% is on trips of 50 miles or less. That means that by far, most of our driving is accomplished locally. Even with today's battery technology, it is practical to design and build a battery-powered car with a 50-mile range. Several experimental models have already been built. Some are being put into limited production. As mentioned earlier, advances in battery technology are certain to extend the useful range as the market develops for electric cars and new-technology batteries are placed into production.

This means that we could eliminate at least three quarters of the gasoline consumption used in private cars if we replaced them with electric cars based on available technology. Since transportation uses more than half of all our oil and private autos use two thirds of all transportation fuel, that would amount to an elimination of one quarter of our total current oil consumption—*if we generate the electricity without burning fuel.*

In our imaginary role of Mr. and Mrs. Average American, we will keep one of our gasoline-powered cars for a while to use on those long trips but buy a new electric car to do all our commuting and around-town driving.

Even the cost to drive an electric car can be less than for a gasoline-powered one. If we drove at 55 mph, the electric car would use about 20 kilowatts of battery power in an hour, which would require 30 kilowatts to recharge. In other words it would use a little more than half a kilowatt hour per mile. If we paid eight cents per kilowatt hour for electricity, that would be four cents per mile for the "fuel." This is comparable to what we would pay for gasoline for a small car if we got 30 miles per gallon and paid $1.20 per gallon of gasoline. In the time frame we have been considering, gasoline will probably cost anywhere from $2.00 to $4.00 a gallon (which is the price of fuel in the rest of the world in 1995). If we consider $3.00 per gallon as a likely cost in this time frame, the mileage cost is 10 cents per mile, or two and a half times higher. Battery life and replacement costs are the key economic questions that can only be answered conclusively with real experience and the passage of time, but to an engineer, the answer is already clear. The days of the gasoline-powered automobile are numbered, *if* we can provide low-cost electricity from a nondepletable source.

In the area of heavy transport, the most promising candidates for electric power are railroads. A few railroads in the US are already electrified, but they are a very small percentage compared to diesel locomotives. In England and Europe many of the railroads are electrified. They developed this way because cheap oil resources were not available and it was more economical for them to generate electricity with coal or stationary nuclear power plants than to import oil. There is no reason, except for the capital costs of conversion, that the US could not change to electricity, *if* we develop nondepletable electric power generation—and particularly if the US were to use the English system with the conductors mounted alongside the rails so there are no visually objectionable overhead wires.

Trucks, buses, aircraft, and ships are a much tougher problem. The most likely solution would probably be the continued use of oil in some cases, synthetic fuels, or in the future, the use of hydrogen made from water by electrolysis. This segment of energy use is probably the most difficult to convert to electricity; however, it represents only about 12% of our total energy consumption, so remaining oil reserves or synthetic fuels should be able to supply the future demands as we continue our search for an alternative solution.

Concepts for the Future

It is likely that as technology is developed and the range of battery-powered cars is extended, long distance travel would become practical. If you drove up to the service station of the future just off the interstate, the "fill-'er-up" request might be a quick exchange of battery packs, with the depleted pack unplugged and slid out onto a service cart and a freshly charged pack slid in as a replacement. After the exchange you would be ready to go again even before the kids are back from the restroom.

Some of the developments being investigated include hybrid systems using flywheels to help store energy for extra acceleration and recovery of braking energy, or small internal combustion engines to work with the battery system for long-range vehicles. In today's cars there is no way to recover the energy expended by the brakes. This energy is lost as heat. In an electric car with the right kind of design, a very large percentage of normal braking energy could be returned to the batteries. Only in the event of very rapid stops would the energy be lost as it is today.

The technological progress that has evolved through the years now makes it possible to develop efficient, reliable, and reasonably priced control systems so necessary for the practical development of electric vehicles.

I have been talking about battery-powered cars because they are the most logical development using what we know today. That does not mean that as the shift to electric cars occurs, the stimulated fertile minds of the next generation will not come up with some breakthrough in the storage of electric power. The economic motivation will be there to find a better mousetrap. I foresee that improvements would evolve rapidly after the switchover to electric cars begins.

One potential alternative energy storage system is flywheel batteries instead of chemical batteries. One company based in the state of Washington is developing such a system. They are projecting ranges of up to 350 miles between recharges with a weight less than a chemical battery. If they succeed it will be a tremendous breakthrough.

As we look into the future of long-distance and heavy highway travel, one of the things that might develop would be the electrification of the interstate highways. This could be done by burying the electric power system in the roadway and having the energy transmitted by induction or radio frequency wireless power transmission to power the vehicles. If this could be done efficiently, then even the trucking industry could use electric drive. This step is not as far off as you might think. Prototype bus systems are already being tested. They transfer electrical energy by induction as the bus stops for passenger pickup.

There would undoubtedly remain many applications that could best be served by the internal combustion engine or gas turbine engines, such as farm tractors that plow the fields and airplanes that ply the sky. However, since much of farming is now being oriented around automated sprinkler systems that rotate about a central pivot and form circular patterns across our great land, why couldn't we have an electric tractor cultivate this circular field using an extension cord from the central pivot for power transfer? Why not indeed? Maybe we could also build an airplane that only had to have enough fuel for takeoff and landing. After takeoff it would climb above the clouds and there switch over to laser-furnished power beamed down from a solar power satellite for the cruise phase of flight. Does that sound farfetched? Yes it does, but a few years ago Abe Hertsberg and a team of students at the University of Washington made a study of the concept and believe it could be made to work.

Another aspect of converting to expanded use of electricity will be the need for more large-scale electrical distribution lines. These lines already exist throughout the country, but they take up large swaths of ground as the towers and wires wander across the countryside. It would be very desirable to eliminate these rather than expand the number. Modern technology may be ready to provide a means to accomplish that goal in the future. We have all seen the announcements of breakthroughs in the development of superconductors that no longer require cryogenic temperatures. If these materials can be developed in sufficient quantities and at low enough cost, they could be used to replace the overhead distribution lines with buried superconductors requiring very small right-of-ways. Even if some refrigeration was still required, they could have better overall efficiency than current high-voltage overhead lines. It is an intriguing idea and one that could have far-reaching implications for our electrical future.

Manufacture of Liquid Fuels

A much more likely course for some of the more difficult conversions will be the use of synthetic fuels or fuels made with the use of electric energy. Hydrogen is the cleanest burning of all fuels, and it can be made from water through the simple process of electrolysis. If you took high school chemistry, you have probably made hydrogen in the lab. Excess power available from solar power satellites during periods of low demand could be used to generate hydrogen. The hydrogen could be liquefied at cryogenic temperatures (minus 423 degrees Fahrenheit) for denser storage or could be absorbed in newly developed solid devices that are proving to be very efficient for storing hydrogen. Some studies have been conducted on using hydrogen gas piped through existing natural gas distribution systems as a replacement for natural gas. This option is possible for the future, but the more likely economical solution will be to use electricity directly since it can do anything that natural gas can do, and with the advent of the modern heat pump, work at very high efficiency.

Other developments are going on in conjunction with ground-based solar power that use the intense heat of concentrated sunlight to produce synthetic fuels or hydrogen without going through the electrolysis process. By the time the satellites can provide large increases in the amount of electricity available from a renewable source the technology to utilize it will be available.

The Changing Energy World

I am sure you are beginning to get the picture of a world that is very different from what we now know. You are correct in that feeling. We are near the end of an era. If we are to be a free, dynamic, and economically expanding nation, we must recognize that change must be made if we do not want to wither away with a dying era. Change is always difficult, but at the same time, it is always challenging and full of opportunities. It will not happen overnight but will cause major shifts in business and jobs. Some businesses will no longer be needed while others will spring up in their places. The internal combustion engine mechanic will have fewer and fewer engines to repair while the electrical and electronics repairman will be in great demand.

As we move into the next energy era we can look forward to the time when electricity will supply 85% to 90% of all our energy use. This will make the energy contribution needed from liquid fuels well within their capability. They could be the products of biomass conversion, synthetic fuels, oil pumped from the ground, hydrogen made from water, liquefied natural gas, or some new source.

Imagine the change in our environment as we convert from burning fossil fuels to the clean nonpolluting energy of the sun. We will be able to eliminate smog, high carbon monoxide levels, high carbon dioxide levels, acid rains, and the industrial haze that covers the states east of the Mississippi River. We will be moving forward with a legacy of clean air for our children. We will also be giving them the abundant energy they will need to exercise their dreams of making a better life without the chains of scarcity holding them back. They will be powered with electricity, the energy form of the future.

8

Our Situation Today

I have discussed the challenge we face as we prepare to enter the twenty-first century. I have told you about the background of solar power satellites and explored our energy heritage. I have reviewed the impact of the 1973-74 energy crisis and what it has done to our country and the world. I have measured the capability of the known energy options against a set of criteria for the future and found there was only one source that has the potential to pass all the tests. So what is the next step? To answer that question it will be useful to look at the energy situation as it exists today.

The profile of United States energy use falls into two major energy segments. One is electric energy and the other the direct use of energy for heating and transportation. Nearly all of the latter is furnished by fossil fuels: oil, natural gas, and coal. However, it is the electricity generation that is important to the future as we look to the development of new energy sources.

In recent years the use of electricity has expanded at about the same rate as the population has grown. The majority of the added capacity has been from coal. This has been accomplished by adding new plants, upgrading old plants, and increased utilization. At the same time nuclear energy has also contributed to some of the increased electrical generation, not from additional new plants, but from increased utilization by operating existing plants a greater percentage of the time. This increased utilization has peaked and net energy production will now decline as older plants are shut down and no new ones are built.

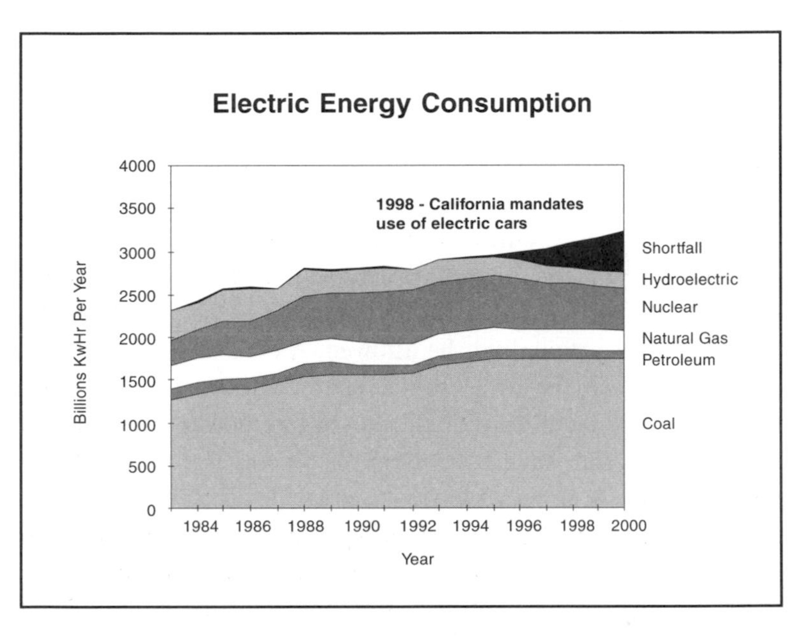

Today, natural gas is the energy source being expanded by the utility industry to satisfy growing demand. Gas turbines are relatively cheap and can be purchased in small increments of capacity without a great deal of local environmental impact. Also, the price of natural gas is low at this time. The utility companies see it as a solution for the next few years, but it will not last indefinitely.

What about the future? There are warning signs. California has mandated that if an automobile manufacturer wants to sell new cars in California, starting in 1998 at least two percent of their vehicles must be *emission free*. This percentage increases with each passing year. For all practical purposes emission free means electrical power. Other states are considering similar legislation. The motivation behind this move is to reduce atmospheric pollution. However, it makes no sense at all to burn fossil fuels in a power plant to power electric vehicles with the intent of reducing atmospheric emissions. As a result there will be immense pressures to eliminate burning fossil fuels to generate electricity. Where will the energy come from to run electric vehicles? The requirement will start small, but it will grow. The amount of energy in the gasoline used to run automobiles is as much as all the electricity currently being generated. This means that electric power generation must *double* just to handle automobile travel without even considering growing demand from an expanding population or replacing the fossil fuel energy currently used for heating.

This is not the only sign of difficulties ahead. The electric utilities are facing an uncertain future when the industry will be deregulated. Competition will be allowed. A consumer will be able to buy power from the lowest cost supplier. Utilities that have high-cost generating facilities will be able to buy power from a lower cost producer and resell it to their customers. The threat of this kind of future is already causing a great deal of concern in the utilities. They are restructuring their organizations, downsizing their work forces, accumulating cash that would normally have been spent building new generating plants, and reducing their costs of operation, all in anticipation of a fiercely competitive future.

If what happened in the telephone and airline industries after deregulation is any indication, we can expect some major upheavals in the electric utilities business. What is likely is the industry will break up into three different kinds of organizations. One will be similar to most current utilities. They will have generating capacity and will distribute and sell the power to the consumer. Another type will become only distributors of power without having any generating capacity. They will buy power from the lowest cost producer. This is how many public utility districts operate today, but the options on where they can buy power will increase. Then there will be the energy supplier that only sells to other utilities in the distribution business. Bonneville Power Administration, for example, does not distribute power directly to consumers, but rather sells power to utilities. In the future they will be joined by major competitors in the business of making money from generating and selling power. The first major move in this direction is coming from industrial companies who are building power sources for their own use and selling the excess power. Many of these facilities use co-generation, where electrical energy is a by-product of excess heat produced for other purposes. The results of deregulation are uncertain, but it is clear that the industry will be in turmoil for some time to come.

The situation today? There is a growing demand for electricity. The utility industry is depending on natural gas for the short term, with a growing awareness of the need to eliminate atmospheric emissions. The low price of natural gas is causing a large segment of the utility industry to turn away from considering any other source of energy because of the tremendous competitive pressure to maintain low energy cost in the short term. The inevitable long-range result will be price increases in natural gas as the demand rises for this finite resource. The utilities have no idea from where new emission-free energy will come, and they are uncertain of their business future. And the Department of Energy has no cohesive energy policy. As one senior DOE official told me recently, "If you review the charter of the Department of Energy and the functions they perform, the last thing you would conclude is that they work on energy programs."

Foreign Nations Are Forging Ahead

During the 1970s the United States was the only nation seriously studying solar power satellites. However, a great deal of interest was stirred up in other countries. While I was the Boeing program manager a great number of people from all over the world came to find out what was happening. This included the news media as well as the technical community. The Japanese sent a whole television crew. The Australians did a TV special on solar power satellites.

One day Lucien Deschamps, a Frenchman, stopped by my office and we spent several hours discussing the concept. When he left I gave him copies of many of our illustrations of various aspects of the concept. I saw him again recently, and he asked if I remembered his visit and the information I gave him. Today he is leading the French efforts to develop the solar power satellite system through a series of progressively more ambitious steps. One of the French initiatives, with the support of the Japanese, is to install a wireless energy link on Reunion Island in the Indian Ocean. It would be from one ground location to another several kilometers away and would be the first continuously operating wireless energy transmission link in the world. France also has plans for space-to-space wireless power transmission tests.

Germany and other European countries are also conducting studies but have no national programs. The former Soviet Union was developing space technologies required for solar power satellites, such as large space structures. They did not have a centralized program, but their technology studies were quite comprehensive. This base still exists in Russia today, though they do not have the financial resources to develop the system. They could be a very effective partner to any country with the necessary capital and desire to develop the system.

India has expressed interest and requested help from the US Department of Energy. DOE told them they had looked at the system in the past, but had no interest in it. An acquaintance of mine from India who works for the United Nations told me the only hope for the developing nations to have enough energy in the future is solar power satellites.

The most significant activity today is occurring in Japan. They started with the "reference design" developed by Boeing for the DOE/NASA system definition studies. The reference design is the satellite configuration developed during the studies that was used to evaluate the concept of solar power satellites against other energy sources. With this solid base they have refined the concept using technology advances made over the years. Some of the advances have come from the US and some are being developed in Japan. One area of intense study is the radio frequency generators for the energy beam. The Boeing design used klystrons. The Japanese are concentrating on developing solid-state devices instead. Their large electronics industry gives them a sound base for this development. They now have an updated reference system definition, and as I mentioned earlier, conducted the first test of wireless energy transmission in space in February of 1993. This test successfully transmitted 800 watts of power from a mother ship to a daughter ship a short distance away. They plan a low-earth-orbit solar power satellite demonstration in the year 2000.

The recent launching of their new H-2 rocket gives Japan their own capability to launch relatively heavy payloads for developmental testing. Though it is not a reusable space transportation system and therefore not capable of launching hardware for an operational satellite, it is still a significant step forward. If Japan and Russia were to join forces with Japan furnishing the capital and electronic technology and the Russians supplying space transportation and space environment technology, they would become an extremely powerful competitor.

The pace of development may be slowed somewhat by Japan's recent business recession, but they still have a stated goal of providing 30% of the *global* energy needs by the year 2040. They know that solar power satellites are the *only* way they will achieve that level of energy production and distribution. It is a frightening prospect for the United States.

What Has Happened in the United States?

During the years since the oil crisis of 1973-74 many ideas have been advanced and billions of dollars have been spent on research, but no permanent solution has emerged. Many politicians have taken the position that if we just ignored the problem, it will go away. But it won't go away.

One of the reasons nothing has happened is because there is not an obvious, easy solution. All of the systems that received extensive funding through the years have either dropped by the wayside or are struggling to survive. When they are measured against the criteria for a future energy source, it is clear why they have not emerged as viable options. If their capabilities had been measured against a set of realistic criteria in the beginning their shortcomings would have been obvious. At that point an intelligent decision could have been made whether to continue development or not depending on comparison to the competing systems. Unfortunately this was not done. As a result, solar power satellites, the one energy source that can meet the criteria, were not developed.

There are several reasons for this. First, it is a high-technology space concept and does not fit the image that people have of an energy source. There are no smoking chimneys or imposing cooling towers or concrete buttresses holding back the water. Rather, it inspires the image of science fiction, rockets, astronauts, high-cost satellites, and unreliability. It is a completely foreign concept to people in the energy business who are used to dealing with labor unions representing coal-dust-covered miners emerging from black holes in the earth, free-wheeling drilling crews coming in from an offshore rig, acres of giant storage tanks around a smoking refinery, super-tankers wallowing across the oceans, and giant cooling towers wafting condensation into the sky.

Second, the system is only cost effective if the satellites are huge, with outputs over a 1,000 megawatts each. Associated with size is the requirement to develop the entire infrastructure to support launching, assembly in space, and maintenance. This means that the initial development bill is very high. And it all must be paid before the first kilowatt-hour of power is sold. It will take farsighted industry and government leadership and support by the people to commit to the massive effort required. That kind of commitment is not easy to make. It is beyond the resources of any one commercial company or even most nations to undertake, so it will either require government support for at least part of the development or a partnership of many companies or nations.

The third reason is the resistance of the currently established energy industry. Oil and coal companies do not want to see a competitive source developed that will eventually put them out of business. This is not true of all companies, as some consider themselves to be *energy* companies and are working to develop alternative technologies that will keep them in business as new energy sources are developed. However, most of the energy businesses in the United States (the largest industry we have) are not about to let a new energy source disrupt their fat, rich, comfortable world without a fight. They will not give up easily. They do not fight the issue in the open, but through their political lobbying base in Washington.

The fourth reason comes from an unexpected segment of our population, but it has been an effective obstacle of progress to date. A small segment of the scientific community sees solar power satellites as a high-technology engineering venture that will disrupt scientific investigation by drawing interest and funding away from scientific programs. These scientists often hold prestigious positions in the scientific community and in government, and their opinions can influence many people, particularly politicians. Some have made uninformed dogmatic statements about solar power satellites, such as "it is impossible to build anything that big in space" or "the cost will be a hundred times higher than estimated." They follow in the footsteps of those who said "man will never fly." Engineers, on the other hand, focus on the physical application of scientific principles. Ideally, science and engineering work in concert to produce new benefits to mankind, but regretably this doesn't always happen.

In December of 1994 I attended a workshop titled "International Space Cooperation: Getting Serious About How," which was sponsored by the American Institute of Aeronautics and Astronautics (AIAA). Two of the five topics were "International Cooperation in Space Transportation" and "Solar Power to Earth." The space transportation topic focused on the need to develop a low-cost transportation system, and the solar-power-to-earth topic focused on what was required to initiate a program for solar power satellites. Attendees at the workshop were senior government and business executives, scientists, and engineers. After the conference was over I wrote a letter to the 60 participants asking for support in developing a low-cost space transportation system and in establishing a program in the US government on solar power satellites to be the focal point for international cooperation.

The replies I received were very supportive, with one exception. The writer—a scientist—reflected the attitude of a small number of scientists opposed to the concept when he wrote: "I cannot concur with your finding that solar power satellites are now practical, if a low-cost space transportation system was available. Your definition of low cost would require a two order-of-magnitude reduction in launch and launch preparation costs. This is clearly not achievable in the near term, nor in the foreseeable future. Even a single order of magnitude reduction is not achievable. . . ."

It is interesting that the part of the system questioned by scientists is the area of cost, where they have the least amount of knowledge or experience. In 1981 the National Research Council (NRC) issued a report titled *Electric Power From Orbit: A Critique of a Satellite Power System.* The report by this prestigious group of scientists was given wide distribution and has often been used as justification for not spending any funds on developing solar power satellites. The critique was made from a scientific point of view and reached eight major conclusions, summarized in the following:

- No development funds should be committed for a decade.
- NASA and DOE should periodically review the system.

- Solar power satellites are technically possible, although costly.
- There are serious doubts about validity of cost estimates.
- Solar power satellites are not competitive with other energy systems.
- Size would strain US resources.
- Questions of international legality and politics exist.
- Development requires worldwide participation.

The critique centers around cost, the size and complexity of the program, and international politics. In the technical area, which is rightly their area of expertise, the scientists agreed it was feasible. The problem is when they criticized the program in the areas where they did not have any experience. For example, they questioned the solar cell cost estimates. Boeing and Rockwell International, the systems contractors who conducted the studies, estimated the cost of the cells based on manufacturing the cells in a production facility. This was an absolute requirement since each satellite would require billions of cells. This resulted in cell cost estimates well under a dollar a watt. The NRC scientists, on the other hand, only considered the cost of cells being manufactured for satellites in 1980 when they were being made in small quantities in research laboratories at a cost of over $1,000 a watt. The scientists simply multiplied those costs by the number of watts needed for a solar power satellite and arrived at astronomical numbers. Based on this kind of analysis their conclusion was that the costs were too high and work should be stopped. Today the solar cell industry manufactures over a hundred megawatts of cells per year and is approaching the estimates made by Boeing and Rockwell in 1980. The cost in 1995 is down to $2.50 to $5 per watt, with the industry projecting $1 a watt in the very near future.

The negative report by the NRC plus the opposition of the Carter Administration effectively stopped all further work. With solar power satellites out of the way, the Department of Energy could continue fusion research, breeder reactor research, and all of its other nuclear activities without the threat of a competing new energy source eliminating their funding. The solar power satellite program has not been able to recover from that devastating blow.

9

A Call to Action for the United States

The question still remains: "What should we do now?"

My answer to that question is clear and unequivocal. *The United States should proceed immediately, with all possible speed, to develop and deploy the solar power satellite energy system as the energy source for the twenty-first century.*

The first half of this book addressed the importance of energy to the development of civilization and the contribution made by each source. Of particular importance was the economic growth associated with the nations that first made use of each new energy source. It also identified solar power satellites as the energy system for the future. The remainder of the book addresses why and how we, as a nation, should go about developing solar power satellites.

The Case for Developing Solar Power Satellites

Some of the very reasons for not developing the solar power satellite concept are also the best reasons to develop it.

First of all, if we were to commit to its development it would give us national purpose. We would no longer be wondering what to do the next time we run short of oil or a megalomaniac threatens to take control of a major oil-producing nation. We would be concentrating on a single common goal—not a generalized wish for energy independence, but a specific solution. It would be a greater task than going to the moon in the 1960s, but it would focus the nation's talents, its energies, and its imagination in much the same way as did that lofty accomplishment. It would challenge our young people to take their place in history building a future for themselves and their children. They would become known as a generation of visionaries who stood at the crossroads of history and chose the pathway of growth rather than stagnation.

It would utilize the talents of scientists, engineers, and companies who have been working on military hardware, which is no longer a number one priority with the ending of the cold war. It would develop a new high-level technological base, which is so important to a highly developed nation like the United States in order to maintain our competitive place in the world economy. It would create a massive number of jobs that would bring growth to our economy.

When the energy starts to flow from the sky it would bring a continuing stream of wealth into our country. We would no longer be dependent on foreign oil. We would no longer participate in the massive exploitation of the earth's resources. We would eliminate the need to burn huge quantities of fossil fuels and thus reverse the deterioration of the earth's atmosphere. It would dramatically extend the life of precious oil for use as a petrochemical and fuel for airplanes and ships, so it could last far into the future. It would build the infrastructure of space development, which would open the space frontier for massive commercial development.

Traumatic changes would certainly affect some existing industries. The coal mining industry would be directly impacted as new solar power satellites came on-line and the old coal plants could be shut down and dismantled. This would not happen overnight, but rather over an extended period of time, giving the labor force an opportunity to acquire new jobs in the expanding economy. Companies would have time to branch out and enter new facets of the energy business as coal mining closed down. Oil companies would also face a shrinking market, but their products would have the advantage of maintaining a viable market for a much longer period of use. Their product base could be modified to encompass the new evolving energy field, which will include solar cells, batteries, and many other components.

Necessity Arouses American Creativity

During the second World War, the nation faced many crises, but two particular stories are worth recounting here. When war broke out, most of our natural rubber supply was cut off. Without rubber, our modern war machine was literally immobilized. Rubber was also needed for automobile tires. The civilian population running the factories would soon be left stranded without the means of getting to their jobs. Something had to be done about the rubber crisis. Several different approaches were developed that might have worked, but the rubber producers were pulling in different directions.

The situation was deteriorating rapidly and a solution had to be found if we were not to go down to defeat by default. Leadership finally surfaced in the form of the War Production Board. They looked at the options and selected Buna-S as the synthetic rubber to be developed. All effort was directed to the development of that one process. They started building production facilities even before knowing all that was required. The entire industry turned around, and within months the problems were solved, production began, and the allied nations rolled to victory on tires built of synthetic rubber produced as a result of a focused national effort to solve a problem.

Another classic example is the Manhattan Project. This project was initiated during World War II by President Roosevelt and was cloaked in the greatest secrecy. Its goal was to develop the atomic bomb. At the time, it was by far the most challenging of all scientific and engineering endeavors ever attempted. It called for scientific invention and sophistication that was mind boggling. The manufacturing accomplishments to separate uranium 235 from uranium 238 were success stories of the highest caliber just by themselves. In addition, new materials had to be developed. Teflon was first made in 1938 and was developed into a useful material by the Manhattan Project. The very best minds of the nation were called to the task. The greatest obstacle was the secrecy lid, which prevented the normal free exchange of ideas so necessary for rapid development. Nevertheless, they succeeded. The world would never be quite the same again, but then it never is after a giant step has been taken. Humanity must continue to move forward if we are not to become a footnote in nature's history like the dinosaur.

"One Small Step for Man . . ."

A more recent example of the dynamic benefits of focusing a country's talent on a specific goal was the history-making event of putting a man on the moon. In 1961, the United States was in the midst of a cold war. While the United States was plodding along on a low-priority satellite launch program, Russia astonished the world by launching Sputnik I. On October 4, 1957, the new frontier of space was opened. While we watched our rockets sputter and burn on the launch pads, the USSR followed Sputnik with more satellites and finally dealt a humiliating blow when Yuri A. Gagarin orbited the earth on April 12, 1961.

Castro was firmly entrenched in Cuba, and the Bay of Pigs affair had left the stink of defeat and disgrace upon us. The rest of the world was asking "What has happened to this once great nation?" We were embarrassed and ashamed. What could be done that would raise us above ourselves and make us forget our failures? What could give us back our pride? What could be done to give focus and purpose to the American people? The cold war was raging, but a hot war was certainly not the answer. It needed to be something that would unite our spirits, stimulate our economy, and allow us to stand tall again and look the rest of the world in the eye and say "Yes, we are Americans!"

In the spring of 1961 President John F. Kennedy rallied the nation to muster its talents, its money, and its people to focus upon a goal. A goal of the highest ideals. An objective that would be very difficult to achieve. One that many thought impossible. An achievement that could make a people proud again. A target that could focus the technical talent of the nation. One which, without resorting to war or war machines, could catapult the nation onto a technology level well beyond any other country on earth. It was just twenty days after Alan Sheppard made America's first brief 15-minute manned flight into space that Kennedy stood before Congress and said, "I believe this nation should commit itself to achieving the goal, before the decade is out, of landing a man on the moon and returning him safely to earth. No single space project in this period will be more impressive to mankind, or more important for the long-range exploration of space, and," he paused, "none will be so difficult or expensive to accomplish."

Engineers and scientists took a deep breath, swallowed hard, and said "Yes, sir." Few believed it was possible in the time available. Many voiced open criticism. But the challenge and goal were clear. The nation rallied. The clock started ticking. No one had a clear understanding of how it was going to be done, only that it would be done.

Teams of contractors were selected, colleges and universities were canvassed for engineers, new factories were built, test sites were activated, Cape Canaveral crawled with workers, and the designers sweated over their drawing boards. Research labs were hard pressed to find solutions to many new problems. Weight was so critical that not an ounce of excess could be allowed. The accuracy of the guidance systems requirements drove the developers into inventing whole new technologies. Workers had to learn how to handle large quantities of liquid hydrogen, which had to be stored at minus 423 degrees Fahrenheit. The term "cryogenics" entered our vocabulary.

Soon the milestones began to tick by, but there were some rough spots. There were test failures. Engines blew up. Propellant tanks burst. Tempers flared over the decision of whether to rendezvous in earth orbit or in lunar orbit. The ultimate agony came on January 27, 1967, with the deaths of Gus Grissom, Ed White, and Roger Chaffee in the Apollo 4 capsule during ground tests. The program was staggered, but a commitment is a commitment and the work pressed on. Nobody said it would be easy.

In December of 1968 we were back on schedule when Frank Borman read a Christmas message to us from aboard Apollo 8 as he and his crew circled the moon. But time for the ultimate goal was beginning to run out and the decisions were getting tougher and tougher. I remember when a three hundred million dollar gamble had to be taken to salvage the Apollo 10 flight schedule.

Part of my job on the Saturn V program was design manager for the fuel tanks on the S-IC first stage. It was five days before the scheduled launch and the Saturn/Apollo stack sat waiting on the pad. The first stage tank for the Apollo 10 mission had already been fueled with RP-1 when an error by a technician at the launch site caused the flow of pressurizing gas to be accidentally shut off. The upper bulkhead on the first-stage fuel tank collapsed. The 33-foot diameter welded aluminum dome had been sucked inward like a deflated balloon—a major disaster that could have lead to a delay of the moon landing by six months, moving the schedule into the next decade. The technician, hearing the crush of the collapsing structure, turned the pressurizing gas back on and was rewarded with the deafening boom of the bulkhead being reinflated. Why it held is a question I still ask myself, but it did. We called together a team of engineers, NASA experts, and an Air Force materials specialist from around the country via teleconference calls. The damage was assessed by working through the night, and tests were run on spare parts. After the options were reviewed, we made the decision to re–pressure-test the tank, with the fuel still on board, while the assembled rocket sat in lonely majesty, bathed in light on the pad.

It was a huge risk. If the tank had burst, the detonation would have been heard around the world. Not the sound, but the doubts it would have raised about America's ability to build reliable launch vehicles and to reach the moon by the end of the decade. The Soviet Union was still in the race. Their rockets had exploded, hidden from the world, but they were still in the competition with rockets even bigger than Saturn V.

It was a tough decision. If we were wrong we would probably not meet the goal of landing on the moon before the end of the decade. On the positive side we could conduct the test with all personnel evacuated from the launch stand so that there were no lives involved in making the test and we could be absolutely confident that if the tank passed a proof test that it would not fail during the launch. Another factor was if the rocket had been disassembled, the delay would have actually cost more than the $300 million loss of the Saturn/Apollo vehicle if the tank had failed. I'll never forget sweating through one of the most traumatic moments of my life. Luckily, as it turned out, we were right: the test was successful and Apollo 10 lifted off on schedule.

Two months later Apollo 11 lifted off with Neil Armstrong, Buzz Aldrin, and Mike Collins aboard. Who will ever forget the thrilling moment on July 20, 1969, when Neil Armstrong stepped onto the moon and said it for all of us. *"One small step for man; one giant leap for mankind."* Eight years and three months had passed since President Kennedy sent man off on the greatest quest in history. Truly an impossible dream became reality.

Our nation was proud again. The world once more knew that when the US rallied to a common cause, we would succeed. Our technology once again leap-frogged to the forefront of the world.

The technological benefits alone have more than paid for the monetary cost of this great human adventure.

The three endeavors cited above all have the same characteristics: a driving national need, a total focusing of effort toward one goal, total commitment of available talent and money to achieve that goal; and, most important, complete success. Only with total commitment can we hope to accomplish our goals, without it we fail. We failed in Vietnam where we went with reservations and limitations toward ends that were not worthy of our dedication. We failed to achieve President Nixon's goal of "energy independence" established after the 1973-74 oil embargo because it was a general goal, without focus, and without commitment to a defined plan of action.

Focusing National Effort—The Key to Success

Solving our energy, environmental, and economic dilemma is certainly worthy of our total commitment. The solar power satellite solution can focus our national purpose on a single effort that will give us "energy independence" by providing a way of directly converting energy from the sun to power our future. It will utilize the technology investment we have already made in space. It will provide economical energy from a source we cannot deplete. It will bring energy to the earth in a form that can be used directly without polluting our environment. It can be expanded to fulfill the needs of all people on the earth as they develop. It will not subject the people of this country to the dragging chains of everlasting inflation driven by fuel costs. It is not a machine of war, yet it would raise our technology capabilities as did the Saturn/Apollo program. It could utilize the capability of the aerospace industry as they turn away from building weapons.

To bring about a decision to develop the solar power satellites will not be easy. I know the frustration and sense of futility, along with those few other dedicated people, as we continue to promote the solar power satellite concept with government agencies who do not want to have anything disturb their comfortable jobs, congressmen who are much more concerned with political maneuvering than accomplishing anything useful, businesses that are only concerned about next quarter's profit and protecting their current product line. *It will require the leadership of the President, appropriate government agencies, and Congress. It will take the backing of the people of America and support from our industries.*

During the period that this option was being actively studied, I was one of several individuals who made many public addresses on the concept and was always impressed by the strong positive response of the audience. There were always many questions about different issues, but almost without exception, the reaction was "Why don't we get on with it?" Why not indeed.

Most people were surprised to learn that there was a possible solution to our energy problem. Unfortunately, the average citizen is not in a position to make the decision to proceed. In the ensuing years little activity in this direction has been carried out, and the concept has drifted away from public view.

Because of the fact that it takes decades of normal development and implementation to bring new energy systems on-line to the point where they can make a significant contribution, it is urgent that the development be started soon, or we will be caught in another round of spiraling costs. It may already be too late for a smooth transition.

The cost of a high-technology energy system, like solar power satellites, breaks down into two major parts. First is the cost of development. This includes the design, fabrication, and testing of all the different elements. Involved would be such items as launch vehicles, space habitats, equipment to perform space assembly, the initial satellite, and the design and assembly of the ground rectenna.

The second cost is the fabrication of additional satellites needed to satisfy the energy requirements of the future. These funds would be provided in a manner similar to any other new power plant purchased by electric utility companies today. The pricing of electric power provides for repayment of these costs plus a reasonable profit.

It is the initial development cost that presents the problem. The cost of developing the technology for the solar power satellite as a power plant is not so much a problem, but rather the infrastructure to launch and assemble it. Much of the infrastructure is unique because it will be located in a remote site. To date there has been no need for a transportation system capable of launching solar power satellites, so it does not yet exist. This is the single greatest impediment to the development of solar power satellites. In the past, costs of this nature were funded by government investment, such as the funding of the railroads as they moved west across the nation. It is not unreasonable for the government to fund the development cost of the required infrastructure as a national investment in our future. The magnitude of the development for the necessary infrastructure, beyond what is being developed by the Space Station, would be considerably less than the Saturn/Apollo lunar landing program.

An important lesson was learned during the moon landing program as the costs were controlled within the original target. This was due to the fact that the original completion schedule was maintained at any cost, and the people working on the program were dedicated to achieving that goal. As a result, solutions were found when a problem developed or another team was brought in to solve it. Time was money. Any significant change in one segment of the effort that delayed another segment meant that huge blocks of manpower were being wasted. A delay in the program of a single day cost ten million dollars. Usually government programs experience dramatic overruns. One reason is due to annual funding restrictions forced by Congress, with the result being programs that are often stretched out much longer than necessary, greatly increasing the total bill. *Actual costs are much easier to control when sufficient funds are provided to maintain the schedule.* This is why the Saturn/Apollo program was accomplished on time and within the original cost estimate.

Reaping the Benefits

If we were to make the decision as a nation to move ahead and dedicate ourselves to developing the solar power satellite as our next major energy source, what would be the benefits?

Let us select the option to pay as we go for developing the space-oriented infrastructure. The commercial utility industry could pay for most of the development costs for the power generation part of the system (the satellite and rectenna) through their contribution to the Electric Power Research Institute (EPRI). Funds for the space infrastructure development could be raised by charging a surtax on imported oil, applying a small tax on energy systems that pollute the atmosphere, and applying some moneys from the military budget.

The government sources of money would actually have a positive long-term benefit. The tax on imported oil would be an incentive to US producers, the tax on polluting systems would be an incentive to expand nonpolluting systems, and the diversion of military funds would help keep the aerospace industry strong and ready if needed in the future for expanded military applications.

Suppose we set a firm time schedule for achieving the delivery of the first useful electric power. The goal of the lunar program had been a little over eight and a half years. A goal of ten years would not be unreasonable for the solar power satellite program.

With those decisions made, what happens next?

Development of the satellite design details would require a large number of scientists and engineers. It would also need technicians, lab workers, manufacturing people to develop the new fabrication techniques, test personnel, and inspectors. Much of the development testing could be accomplished on the ground by testing a ground test prototype of the power generation system and wireless power transmission from one location on the ground to a receiver a short distance away. The solar technology would be based on the arrays developed for terrestrial solar systems.

The major task during the development phase would be a new fully reusable space freighter to replace the costly Space Shuttle. Other space transportation vehicles for moving parts from one orbit to another would need to be designed and built. These projects would require a work force similar to several modern airplane companies.

The design and development of the assembly base to be operated in space would require a new breed of workers. Skilled laborers, like those needed to build the pipeline in the Alaskan North Slope oil fields, would have to learn how to build giant structures in space using robotic assembly tools. Development of the ground receiving antenna would require a large number of people familiar with heavy construction, earth moving, and field assembly.

As the program moves from the design and development phase into manufacturing of the initial satellite, the majority of jobs would be involved in building and operating the space transportation system, fabricating the components and subassemblies for the satellite, and constructing the ground receiving antenna. Only a relatively small crew would be required for assembly in space. When I say small, that is only relative to the total work force required, but quite large by any space operations we have today. The total number required will be dependent on how many will be required to monitor the robotic assembly machines and to handle cargo transfer between the heavy lift launch vehicles and orbit transfer vehicles carrying cargo from low earth orbit to geosynchronous orbit.

This project would grow to provide hundreds of thousands of jobs in many different disciplines, encompassing the entire spectrum from highly educated scientists to hourly laborers. The number of jobs generated in periphery fields to support this work force would reach into the millions.

Even though the basic technology required to develop the satellite and its support systems is known in some form today, the application of the technology into an economically successful commercial power plant would require extensive development efforts. This effort would be directed at achieving total understanding of principles, new and simpler techniques for applying them, low-cost automated manufacturing approaches, and extremely high reliability.

For example, this would include the development of lightweight, low-cost, high-efficiency, and long-lived solar cells. Much of this technology has already been developed for terrestrial solar power systems. The new requirement will be to apply the unique criteria associated with the space environment. This technology would in turn make solar cells for terrestrial use much more affordable and practical in many locations. Solar cell power modules used to supply the power for industrial developments in space would be very economical.

Development of the wireless power transmitter would elevate the technology available for radar and other wireless systems to staggering heights. Power switching systems and power processors would reach new levels of simplicity, low cost, and reliability. Routine operation of reusable space transportation vehicles would necessitate development of new and better high-temperature materials, long-life engines, lightweight reusable cryogenic insulation systems, and expanded knowledge of how to live and work in space.

These are but a few obvious examples. If our experience on the Saturn/Apollo program was typical, the real technology benefits to the country—other than the accomplishment of the goal—are often surprising and unexpected. Who would have thought that the requirement for reduced weight and increased capability needed for electronic systems would have opened the way to $10 pocket calculators and personal computers? Even the microprocessors that control so much of our machinery, including our automobiles, are a result of these developments.

The Saturn/Apollo program provided much of the basic knowledge that supports our modern military activities as well. It displayed to the rest of the world our technological capabilities in a clear and dramatic way. The solar power satellite can do the same and even more effectively.

The real potential, however, is the ability to add generating capacity as the demands for energy grow. After meeting new energy requirements we could start replacing the existing fossil fuel plants and obsolete nuclear plants. A large percentage of the current power plants in the country are wearing out, and maintenance costs are accelerating as they reach the end of their useful life. They could be replaced with solar power satellites, thus eliminating the demand for fossil fuels as our major energy source and starting the process to clean up our atmosphere. Once this is done, a more natural growth can occur. With the availability of ample low-cost electricity, the move could be made to replace a large share of the transportation requirements with electric power vehicles as well.

With abundant, low-cost, pollution-free electricity, we would be able to build giant desalinization plants to make fresh water from the sea and eliminate water shortages in much of our nation and the world. In other areas, far from the sea, we could use the energy to recycle waste water to high purity and use it over and over again to supplement nature's cycle.

Freeing the People of the World

As great as the benefits are for the United States, much of the rest of the world has even more to gain. Without sufficient energy to provide the necessities of life, people in the developing nations have no hope of improving their lives.

Many of these people are shackled with the bonds of poverty. Bonds stronger than prison bars and more binding than oppressive governments, for even if prison doors are thrown open and governments allow freedom of choice, what good is freedom if there is nothing to eat, no roof to give protection from the elements, no money or possessions? Survival, by necessity, then becomes the most basic human drive. When the problem of seeking food and shelter dominates all effort, freedom is only a word—without meaning—to people who are starving.

In our crowded world where it is no longer possible for everyone to grow or gather their own food and build their own shelters, impoverished people are faced with two possibilities if they are to realize any kind of freedom.

First they look to the prosperous people of the world and ask for help. If that help is a gift of food or temporary shelter it is soon gone and with it their fleeting freedom. At best a transient break in their lives, the gift does little to help them achieve permanent relief from the bonds of poverty. Even with the best intentions, the wealthy people of the world cannot give enough to meet the demands of all the poor; sadly there are too many.

The second option is much more difficult, because it cannot be given as a lasting gift—it must be earned. Political freedom is often won in the voting booth or with revolution, but such freedom does not guarantee that there is enough to eat and sufficient resources and time to make freedom of choice possible—that can only be achieved by increasing productivity. This requires ample energy and the moral commitment of the people to put it to work.

When there were but a few people on the earth, nature supplied all the needs of everyone directly. There was abundant food, water, and raw materials for all their uses whenever it was needed. As mankind multiplied there was still enough for everyone, but it became necessary to cultivate the fields and raise livestock in order to have a sufficient supply. As human knowledge expanded, so did people's ability to enhance their surroundings and to improve their standard of living. More time was available after performing the basic needs of survival and therefore freedom of choice became a reality.

On the other hand, for the impoverished peoples and nations who did not increase productivity as their population expanded, life became more desperate. They could no longer gather food and water and fuel in sufficient quantities to sustain everyone. They are the truly enslaved and can only hope to experience true freedom in this world through learning and application of how to use energy to multiply their abilities to provide food and lodging beyond the basic need for survival. Only then will they have the time to know they are free.

The greatest gifts that affluent people of the world can give to free the impoverished are knowledge, tools, guidance, and the *energy* to make it all possible. For in the words of one university professor, *"Freedom is a society of plenty."* Only ample energy can make that possible.

10

The Path to Two-Cent Power

The subject of economics is probably one of the most discussed and controversial subjects known to man. This is true because every human being on earth today is a practicing economist with their own ideas and views. There are very few decisions we make in our daily lives which do not consider cost to some degree, even if it involves bartering instead of money. In its simplest form, it is a question of whether we have enough money in our pocket to buy today's newspaper, or in a more fundamental society, whether an exchange of a bunch of bananas is worth a t-shirt. As we move up the scale of economic decisions, the choice of personal transportation stands out—do we buy a new car, and if so, which one? Or, for a laborer in Beijing, China, can he afford to buy a bicycle. In India, it may simply be a decision to buy a new pair of sandals.

Further up the economic ladder comes the selection and purchase of a home. This factor looms very large in our personal lives. It is not a decision made hastily. We consider many things: its cost and whether we can really afford it, how much we will have to give up in other ways, how badly we really want a family room, what type of heating it has, whether it's in a good neighborhood and matches our lifestyle, how long we plan to live there, whether it will be easy to maintain, what the taxes and insurance are and will it appreciate or depreciate with time, and will it be easy to resell if we have to move. What it boils down to is whether we can afford it and whether it's the place we want to live and spend our brief time on earth. Buying a new home is probably the single most important economic decision we make in our personal lives, yet to how many of the above considerations can we actually give an absolute monetary value?

The critical cost factors can be evaluated with great accuracy, such as the down payment and the monthly payments, including tax and insurance, as well as the likely utility costs. In essence we ask ourselves whether our income-generating capability is sufficient to meet the costs. If the answer is "yes," then the real decision is made on the other points. The cost issue acts only as a sieve. After that, other considerations dominate and theoretical economics become totally entwined with human behavior and emotions.

When we make our decision on a home purchase, we need to know the costs with reasonable accuracy before we can make a decision based on the remaining features. The same is true of a grand venture such as the solar power satellites. In this case there are two types of cost we must consider.

The first is to estimate what the development costs will be. These are the costs incurred prior to the construction and operation of a system. They are often referred to as front-end costs and are the most difficult to estimate, since the risk is usually high. In commercial businesses, amortization of the development costs must be spread over many production units, and it is not always certain how many units will be sold. With government involvement, development is usually considered to be a national investment, without attempting to reclaim the cost from direct sale of the product. This is true of military hardware and most government-sponsored research and development. In the past this has included aeronautical development, nuclear energy development, medical research, and many others.

The second cost category is how much each unit will cost as it is produced. These costs are generally easier to predict accurately if we can define the product with reasonable detail. In the long run, these costs are by far the most important.

The issue of cost of a large new energy source is important to us as individuals since those costs determine the cost of energy that we buy and use in our daily lives. They are also critical to the general economy that provides us with jobs and the opportunity to improve our lives. If they are too high the cost of energy will be too high and there is no reason to make the investment. So the first issue we must consider is what will be the cost of energy for the consumer if it is generated by solar power satellites compared to other sources. It must be much lower in order to win our support to make the huge investment required to develop the satellites needed to gather solar energy in space.

To make this initial comparison I will use the cost estimates generated by my Boeing team for the NASA/DOE studies, which were based on a satellite that had an output of five thousand megawatts. The cost estimates to produce a single satellite was $12 billion. This cost was developed in 1979 dollars, and includes the cost of the satellite, the ground receiver, and the cost of transporting the satellite hardware to space. The $12 billion would escalate to $24 billion in 1995 dollars.

These numbers do not include research and development costs nor the cost of the infrastructure required to support satellite development. These two items will be discussed separately in order to keep the comparison as consistent as possible. They will be treated as a national investment in the same way that nuclear power was developed.

How do Solar Power Satellites Compare to Other Systems?

There are several ways to compare costs of different energy systems, and most of these comparisons are slanted toward the business and investment community and how they make investment decisions. In the electric energy generating business this approach often results in minimizing investor risk at the expense of the consumer and results in higher costs of electricity.

The reason for this is that the cost of fuel is seldom included in any comparisons made by investors. Fuel does not represent a capital cost and therefore is not included in the original costs that must be financed to build the plant. Fuel costs are routine month-to-month operating expenses that are passed on directly to the consumer, so investors do not consider that it has to be counted. However, for one coal plant with an output equivalent to a five-gigawatt solar power satellite, the fuel cost will be between $40 and $250 billion over a 40-year lifetime. The range depends on whether there is any inflation and the rate of inflation. It seems to me those are pretty big numbers to ignore, particularly when they will ultimately be passed on to us, the consumers.

Another reason fuel costs are ignored is because most current economic theories use discounting analysis, levelized costing, or present value factors to make economic decisions. All these theories attempt to place mathematical formulas around all criteria with a cost equivalent number assigned. No value judgments are allowed. Based on these formulas the economists say that a dollar in the future will be practically worthless. Or conversely, we can afford to spend $10 to $100 twenty years from now in order to save one dollar today and anything that doesn't produce immediate profit is not a good investment today. Lucky for us our ancestors did not think that way.

When these economic theories are used, only the money spent today is evaluated in the equation; money spent for fuel in the future does not affect the result. They are nice theories, but like so many theories trying to explain complex interacting relationships, they disintegrate in the face of reality. They fail to consider the value of investing in capital expenditures for future benefits.

Some of the economic woes in the United States, besides those being driven by the high cost of energy, were brought about by this economic philosophy. The steel industry is a case in point. Through the years they failed to make the capital investments that would keep them competitive with the Japanese or Europeans, ending up with obsolete plants and processes that were labor and energy intensive. When difficult times came, the bottom dropped out for them.

In our comparison I will not use any of these theories. Ours will be a straightforward economic approach to compare the costs of electricity using the same method as for a typical state-regulated utility company. The cost will include the cost of buying a power plant, operating and maintaining it, and also the cost of fuel to provide the energy. The cost of buying the plant includes the cost of borrowing the money and paying interest over a 30-year period as well as the cost of taxes and insurance. All of these costs plus a profit margin are then passed on to the consumer. The plants in our comparison will be sized to have the same total output capacity over a 40-year life span. Forty years is about the maximum life for a typical fossil fuel generating plant. The life of a space solar power plant will be much longer than 40 years because of the benign environment of space, but for the comparison we will consider the first forty years.

The cost of operations and maintenance has become a major factor in some power plants, for it is in this category that the cost of pollution controls are felt. The cost of chemicals and exhaust scrubbers to control emissions from coal plants is now quite large and growing because the allowed emissions have been greatly reduced to limit atmospheric pollution.

In the United States today by far the largest source of electricity comes from coal, which accounts for 56% of the total. Nuclear now provides 22%, with hydroelectric accounting for 9%, natural gas 9%, and oil at only 4%. Others such as geothermal, ground solar, and wind turbines are so small they aren't considered to be significant.

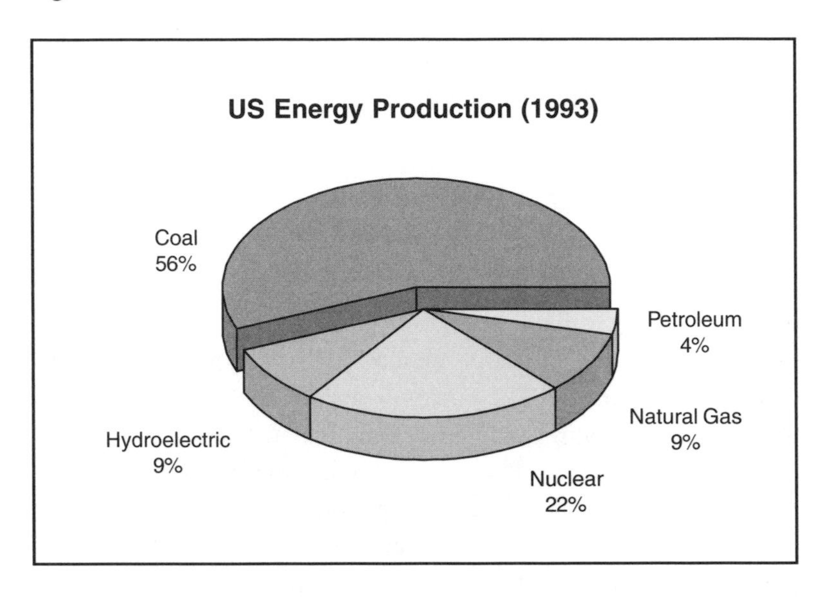

Of all the sources listed above, hydro provides the lowest cost power. Unfortunately we have already built dams on most of our rivers so there is little chance of adding significantly more hydro power. The only exception is the state of Alaska, and it is far from the centers of demand. The last new nuclear plant was ordered in 1978 and none are under active construction at this time. The remaining production is from fossil fuels being burned in thermal power plants.

I will use coal in the comparison because it is the largest source and the lowest cost next to hydro. Coal power plants have been in existence since the nineteenth century, and we would expect that their costs could be accurately predicted. However, there are several extenuating circumstances that cause the costs to vary widely from plant to plant. Four factors come into play that raise havoc with the estimators. First, environmental requirements are a new consideration and have a large impact. The second is site location and regulatory delays. The third factor that estimators have found difficult to project is the future rate of inflation and the cost of labor disputes. The fourth is mismanagement, which has been particularly rampant in public power systems. As a result, the cost of coal plants built during the last decade varied by a factor of three in their cost per unit output, depending on their location and the various impacts of the factors mentioned. If one were to choose a representative average cost figure, it would be about half of what a solar power satellite would cost per kilowatt of generating capacity. Cost per kilowatt of generating capacity is a common and useful way to express the capital cost of a plant because it ignores the unit capacity of individual plants and places all systems on a common comparison basis.

A typical investment comparison would stop here. The fact that the capital investment in a coal plant is half the cost of a solar power satellite would make it the clear winner in an investor's mind—particularly since solar power satellites have never been built and therefore the cost estimates are very suspect to the investor.

As hard as it is to pin down the capital cost of the generating plant the real culprit in the energy cost comparison is the cost of fuel. Coal is particularly difficult to specify. Not only is there a variation in quality due to very large differences in the heating value of different coals, but impurities cause major impacts on environmental emissions. Transportation costs also vary with the distance between the mine and the power plant. For the comparison, I have used a median Northeast cost for coal delivered to the plant. The problems do not stop here, however, because the plant must use fuel over all the years of its existence, so the cost must reflect the effects of inflation, the effects of regulatory changes, and the ever-increasing difficulty of mining coal as it is gradually depleted. Recent history suggests that a real cost escalation of over two and a half percent a year can be expected from regulatory changes and increases in mining cost. I have also used 3% annual inflation as the lowest we could possibly expect over the long term.

There is no fuel cost for solar power satellites so it avoids all the uncertainties of the escalating cost of fuels, and its only costs are capital costs and the cost of operations and maintenance.

Now all we have to do is select a time frame. Since solar power satellites would be a solution for the twenty-first century, I will start the comparison in the year 2000. The electricity cost from a new coal-fired generating plant placed in operation in the year 2000 would be 18 cents per kilowatt hour for that year. Nine cents of that is capital-related cost, two cents of that is required for operations and maintenance, and the remainder (seven cents) is fuel cost. In the same year electricity from the satellite would cost 15 cents per kilowatt hour. Nearly the entire amount is capital-related as operations and maintenance are only a fraction of a penny per kilowatt hour and there is no fuel cost.

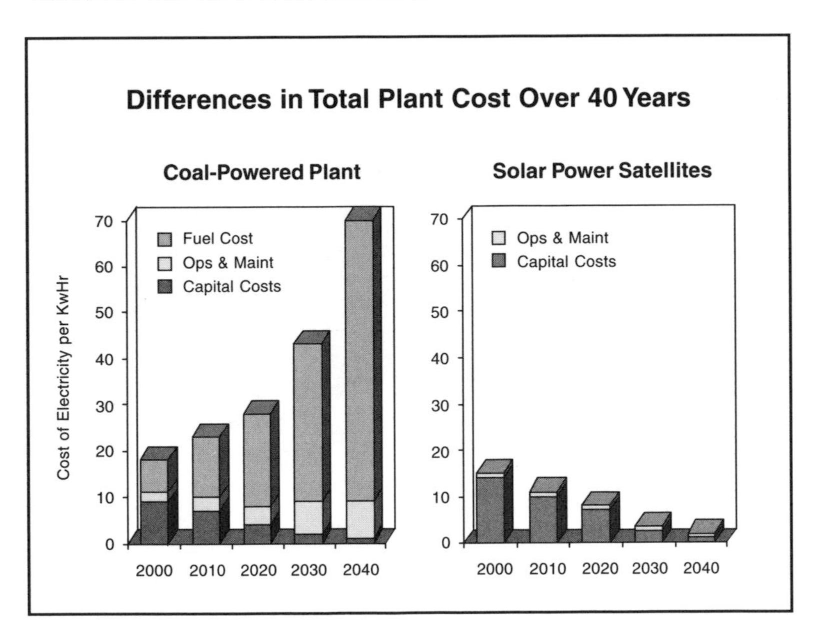

At the beginning electricity costs are similar; however, they start to diverge immediately. By the year 2010 cost of electricity from the coal plant has risen to 23 cents per kilowatt hour while the cost for the solar power satellite has diminished to 11 cents as the original investment is being paid off. Capital costs are also diminishing with the coal plant, but cost of fuel and cost of operations and maintenance escalate with inflation.

By the year 2020 the difference has significantly expanded. Coal-generated electricity cost is increased to 29 cents per kilowatt hour while solar power satellite costs have dropped to 8 cents. Escalating fuel cost is the main cause, but the cost of operations and maintenance is also escalating for the coal plant as emission standards are tightened.

In the year 2030 all the capital costs will be paid off, but this has little effect on the cost of electricity from the coal plant as the cost of fuel and operations and maintenance are the overwhelming cost elements, raising the cost to 43 cents per kilowatt hour. In the case of the solar power satellite the cost of electricity has dropped to 3½ cents, and with the final capital cost repayment the price of electricity drops to under two cents per kilowatt hour by the year 2040. And a large part of that cost is *profit*.

All capital costs will be paid off in 2030 and yet, due to rising costs of fuel and operations and maintenance, by the year 2040 the cost of electricity from coal has sky-rocketed to 70 cents a kilowatt hour and the plant is worn out. In the meantime a solar power satellite would be producing power at less than two cents per kilowatt hour and it should be able to do that for many years beyond the life of a coal plant. With proper maintenance its life can be indefinite, just like a hydroelectric dam.

Imagine the impact on the average household in the year 2040 if they used 3000 kilowatt hours of electricity a month? With coal used to generate electricity they would pay $2100 a month. With solar power satellites the bill would be less than $60 a month. And that is only for their home. When we consider the energy cost impact on all of the things we do and buy each day, the difference will be poverty versus wealth.

Energy Costs Can Be Immune to Inflation

When we look at the potential savings with solar power satellites compared to coal, the numbers become immense. The savings in energy cost from one solar power satellite over a 40-year period would be more than $300 billion, if we had only 3% inflation. If we expand that to represent one half of the current US electrical generating capacity, the potential savings becomes about $22 trillion in a 40-year span with a total investment of less than $2 trillion. After 40 years the savings would grow even larger each year.

One wonders what will happen to the economy in the future if we continue to spend an ever-increasing portion of our wealth for energy. The money we will save by developing solar power satellites can reverse this trend. It will provide a healthy economy that will enable us to pay off the national debt and create a source of unending wealth for all Americans. Even if the capital cost was underestimated by a factor of two or three times, the ultimate cost advantages would not be significantly affected because of the huge financial leverage of the satellite system. This is true because there is no fuel cost to continuously escalate energy costs.

With the satellites, energy would no longer be one of the causes of inflation; instead it would become one of our most effective tools for fighting inflation. This phenomenon is not just a flight of fancy. We have the excellent examples of the hydroelectric dams. Many were built in the first half of the century and are now paid for. The cost of power from these plants is only a fraction of a penny per kilowatt hour. The mortgage is paid off and the water keeps flowing. Our forefathers made capital investments in the future that are paying off handsomely today. Unfortunately there is no more room for big hydroelectric dams. We need instead to build solar dams where there is nearly limitless room and no limit to the energy flow. Solar satellites could be called solar dams with the characteristics of being essentially immune from inflation and without the capacity limitations of hydroelectric dams.

With their promise of abundant, low-cost, clean energy there is only one sensible choice when we are faced with the decision to make a capital investment in the future and build solar power satellites. There are those who will want to do the expedient thing, which is nothing. But can we afford to do nothing? We will be forced to pay trillions of dollars for fuel that will be burned and gone forever as it pollutes our atmosphere. Future generations will judge us harshly and with justification if we do not step up to our responsibility to create a foundation for their future. We had it soft and easy in the golden age of oil. Are we going to be too flabby to make hard decisions that can overcome the difficult times we are experiencing now, which will only become tougher in the future?

The evidence is there, the technical ability is there, and the funding capability is within reasonable and achievable levels. We can afford the down payment, we can afford to make the mortgage payments, and the total costs are lower than other sources. We cannot plead ignorance. We must make our decision to make the capital investment in our future based not only on projected costs, but also on value judgments.

Does your value judgment say that solar power satellites are the answer for the fourth energy era?

11

Costs are Based on Sound Estimates

The cost of electricity generated by solar power satellites compared to coal reaches to the heart of the reason to develop solar power satellites. We cannot ignore the immense economic benefits, not to mention the environmental necessity to stop polluting the atmosphere and cease further accumulation of deadly wastes from nuclear power plants.

The real issue is the validity of the cost estimates that make up the basis for these cost comparisons. Even though estimating is part fact and part guesswork the estimates must be close to reality or the great economic advantage is lost and the system will not be widely deployed. Many factors go into making cost estimates, including related experience and the quantity of parts being manufactured. When a new venture is being developed it is often impossible to compare everything to previous experience, but by comparing some we are able to predict the remainder from similar experiences of the past.

The solar power satellite is simple in concept, but is unique because of its huge size. Because of its simplicity it has only a limited number of different kinds of parts. Its size is the result of assembling a vast number of smaller, identical parts together. For example, all of the solar cells are the same, all of the radio-frequency generators are the same, and most of the structural pieces are the same. As a result of this characteristic most of the cost of the satellite hardware is dependent on determining the cost of items like solar cells that will be manufactured in mass-produced quantities, even for the first satellite. Each satellite would have over half a billion solar cells, more than five times current annual world production. To reach this production quantity will necessitate factories designed for mass production. As we know from experience, the more parts produced, the lower the cost.

In this chapter we will consider some of the critical elements and why their cost estimates are reasonable.

Previous Space Investment

In 1961 when President Kennedy announced the goal of sending men to the moon we were but three years into the space age. The famous rocket designer Wernher Von Braun had developed an idea of how it could be done in 1953. Even Jules Verne, the great visionary and author, wrote a science fiction story near the turn of the century about man's journey to the moon. But the actual amount of serious study of the subject was very limited. A small group from NASA had been working quietly for a couple of years, but that was about all. None of the launch vehicles, spacecraft, or lunar landers were developed, and the space industry base was only in embryonic form. The massive test and launch facilities did not exist. The only large booster under development was the much-too-small Saturn I, and space engineering experience was mainly with military missile systems. In fact, the starting base for such a massive technological undertaking was shaky indeed.

The list of different components, subsystems, and vehicles that had to be designed, developed, and tested to form the complete Saturn/Apollo program staggered even the most optimistic mind. Congress wanted to know what it was going to cost before they would authorize the initial funds. NASA pieced together the best estimate they could based on limited experience and multiplied that by an uncertainty factor to come up with a cost. Their projected program expenditures were estimated to be $24 billion in 1961.

In the ensuing years, as the program evolved, a major new industry was developed. A sprawling, empty, tank assembly plant at Michoud in New Orleans, Louisiana, was gutted, rebuilt, modernized, and expanded to assemble the giant Saturn boosters. The swamps and bayou country near Pearl River, Mississippi, soon throbbed to the mighty roar of stage test firings conducted in the new test stands squatting near the water's edge. In Huntsville, Alabama, giant test facilities grew out of the ground, changing a quiet southern town into the "place where space begins." In Clear Lake City outside of Houston, Texas, NASA brought together the finest team of space engineers and scientists ever assembled. The largest building on earth rose out of the Florida sand to house the coming monsters. Launch complex 39, with its giant crawling transporters, was made ready. In the aerospace companies, engineers learned a new language—no more was it "dihedral" and "wing sweep angle"; it was now "delta-V, thrust-to-weight, mass fraction, staging velocity, insertion velocity, perigee, apogee, main engine cutoff, ISP, chamber pressure"—a whole new vocabulary. We had to learn to think in thousands of feet per second, not hundreds of miles per hour. We learned from a few old-timers in the space business—if you can call people with two or three years' experience "old-timers"!

New trails were blazed. Out of the colleges came bright-eyed, eager young engineers drawn by the challenge, refusing to believe that it could not be done. This new breed was joined by others who came from designing steam boilers to designing propellant tanks. Factory people came from everywhere. The task they undertook was clearly the largest and most technically complex ever attempted. Starting from an ill-defined base and inventing the needed technology these young tigers were dedicated pioneers at the leading edge of man's knowledge in the summer of 1962. Yet, the task was accomplished at a total cost very close to the original estimates. If we were to consider the $24 billion spent on the moon program in today's dollars, the cost would be approximately $120 billion.

Technology Starting Point

The technical challenge of the solar power satellite is comparable to the Saturn/Apollo moon program in complexity and development magnitude, except for two distinct differences. We understand the overall satellite system much better than we understood the Saturn/Apollo system at the beginning of the moon program in 1961—we are not starting from scratch. More important, we now have nearly four decades of space experience behind us. All the technology developed by the lunar program, the shuttle program, the National Aerospace Plane research, numerous satellite programs, terrestrial development of solar cells, development of the Space Station, and all of the big radar systems give us a solid base on which to build.

Space Transportation—A New System is Required

When the United States' first successful satellite hurtled into space on top of a Jupiter C rocket in January 1958, the cost for delivering that payload to orbit was in the hundreds of thousands of dollars per pound. With increased knowledge and abilities, we moved on to bigger and better rockets. Even though we were still throwing away an expensive rocket, the cost of launching Skylab on top of a Saturn V was on the order of one thousand dollars per pound.

Space Shuttle with its promise of reusability was a big new step. We only throw away the external fuel tank and some of the hardware of the solid rocket motors on each flight. The shuttle orbiter, with the expensive engines, electronics, and controls, returns intact to the earth and lands at the launch site like an airplane. It is flown again and again. This is the first generation of partially reusable launch vehicles. Unfortunately the promise was greater than the reality and it has not achieved its launch cost goals, primarily due to politically driven decisions and poor management during development.

The National Aerospace Plane was supposed to fly single-stage to orbit (SSTO) and back with a fully reusable vehicle. This research program used a ramjet/scramjet airbreathing engine and made some major steps forward in the development of high temperature materials, reusable cryogenic insulation, and hypersonic aerodynamics before funding limitations and technical problems with the engine lead to its termination.

There have been some experiments with a single-stage-to-orbit reusable rocket-powered vehicle that looks promising. The first test for maneuvering and landing has already been completed, but government funding is drying up. That is too bad because in a larger size it could be a serious candidate for launching the satellite hardware.

A different concept, developed during the studies in the late 1970s, used a two-stage, winged, fully reusable, vertical takeoff, rocket-powered vehicle, with both the booster and orbiter returning to the launch base to land horizontally like the Space Shuttle. This seemed to be a good concept at the time and is still the most likely candidate for launching heavy payloads.

Even more intriguing, particularly for lighter payloads and as a personnel carrier, would be a large airplane capable of Mach 3 speeds to carry aloft a rocket-powered orbiter to be launched at Mach 3. Both the jet engine–powered airplane and the winged orbiter would return and land at the launch site, with no hardware discarded.

Since the opening of Russia to the outside world, their technology is also available. Engineers who have visited Russia and seen their rocket engines are very surprised at the level of technology and efficiency they have reached. In many areas it is considerably ahead of what has been has been done in the West. The use of Russian-designed and built engines could be a major advantage.

The stage is set by all this activity to take the final step to a fully reusable space freighter and personnel carrier. By not throwing away expensive hardware it will be able to provide the low cost necessary to build economical commercial solar power satellites. Whatever the final design it must incorporate the three basic principles essential to all good transportation systems.

First and foremost in importance is minimal maintenance. This certainly means they can no longer be thrown away and replaced after each flight.

Second, it must be able to carry large payloads and have convenient loading and unloading capabilities, using pallets or cargo containers similar to containerized ships. The use of pallets or containers will expedite loading and unloading of launch vehicles in the same way it did for ships and railroads.

Third, whatever the design, the vehicles have to be able to fly over and over again. A typical airline keeps their airplanes in the air more than 10 hours a day, year after year. They don't make money when the planes are on the ground. A space freighter must have similar capability.

A system designed and operated to these requirements will become a mature transportation system like airlines, railroads, ships, or the trucking industry. The cost of operating mature transportation systems is a function of the cost of their fuel. Operating cost is typically between two and five times the cost of fuel. Today the cost of space transportation is well over a thousand times the cost of fuel. When there are frequent flights and nothing is thrown away, space transportation vehicles will join the ranks of other mature transportation systems. Their operations cost will certainly fall below ten times fuel cost and eventually reach no more than five times the cost of fuel.

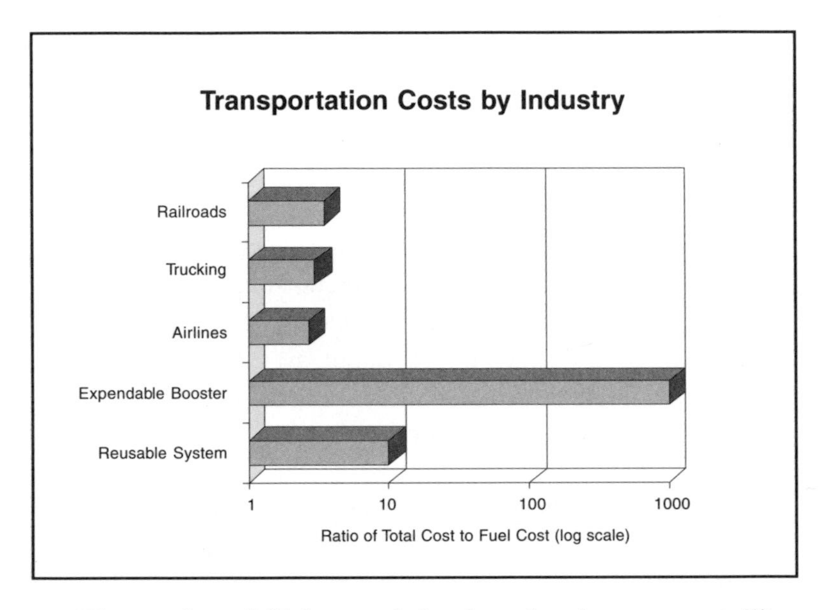

The number of flights needed to launch solar power satellites will require the development of a new reusable space transportation system, but this is not the only program that needs low-cost space transportation. The growing use of cellular wireless telephones is leading inexorably toward space-based worldwide systems. Motorola is already working on their Iridium system, and Teledesic is proposing a more aggressive program that would use 840 satellites. Space travel for the average citizen is the goal of one new entrepreneurial company. As a result of these and other requirements, solar power satellites is not the only program that must pay for the development of a new space transportation system. The cost of this significant part of the infrastructure can be spread between several users.

Solar Cells—Becoming a Mature Industry

Solar cells are another example of space-related development. They are the core of the solar power satellite concept, so their cost is pivotal to the success of the system. In 1980 the uncertainty of their projected cost was one of the reasons for terminating development of solar power satellites. Most of the United States satellites launched over the last 35 years have received their electrical power from the sun via photovoltaic devices called solar cells, which convert sunlight directly into electricity. These cells usually were made from silicon and were very expensive because of the laboratory processing for the small numbers used. Through the years, their efficiency has been steadily increasing, and in recent years, the impetus to develop low-cost solar cells for terrestrial use has greatly increased research in the field.

In the spring of 1994 my colleagues and I conducted a survey of the United States solar cell industry. We visited nine manufacturing facilities, three national laboratories, and four utilities to assess the progress made over the last fifteen years. What we found was very encouraging.

Current world production has grown to about 100 megawatts annually; these are principally terrestrial solar cells of varying types. This level could not support a solar power satellite program at this time, but none of the people we talked to were intimidated at the prospect of the industry as a whole being able to expand production to the volumes needed for a solar power satellite program. The only comments in this regard were that "given sufficient time and sufficient demand" the accumulation of the necessary tooling and the development and use of better mechanized production facilities would certainly take place.

Many photovoltaic manufacturers are building solar cells and arrays using many different technologies. In addition to crystalline silicon and gallium arsenide cells that are the close relatives to transistors and computer chips, technologies now include thin film cells of several different materials. These cells are made by depositing a very thin layer of photovoltaic material on a substrate of plastic film, metal foil, or thin glass. While not as efficient as crystalline cells, they can be lightweight and use very small amounts of expensive materials and have the potential of being very inexpensive. The two most promising thin-film cells are copper indium diselenide and cadmium telluride.

However, crystalline silicon still has the largest market share of commercial production. It has held this position based on improvements in manufacturing techniques and steadily increasing efficiencies. Today a typical commercial silicon solar cell is a thin wafer four inches square that is either a single crystal or polycrystalline. The wafers are made by cutting a thin slice from a silicon ingot. These single crystal ingots are made by an interesting process called a Czochralski pull where a single very large crystal is formed by slowly pulling a rotating plunger out of a bath of molten silicon. As it emerges the silicon solidifies in the form of a single crystal. Polycrystalline ingots are made by casting the molten silicon in a mold similar to the way iron is cast.

Dramatic advances in low-cost production techniques have been made in processing these large ingots, reducing the cost of commercial solar cells to a range of $2.50 to $5.00 a watt. With the terrestrial market continuing to expand, manufacturers are projecting $1.00 per watt in the near future. In 1980 we projected cell costs would reach 25 cents a watt in 1980 dollars when the production rate was great enough to build the first satellite. This would be about 50 cents a watt in today's dollars. There is little doubt that with a ten- to hundred-fold market expansion from building solar power satellites the original cost target of 50 cents a watt will be achieved.

The situation is even better for cell efficiency. In 1980 we were projecting silicon solar cell efficiency would be 16.5%. Today silicon cells can be purchased that are over 20% efficient, and with concentrators to intensify the sunlight their efficiency reaches 26%. Current production efficiencies for most silicon terrestrial solar cells are in the range of 12% to 14%, but are expected to increase as new advances are incorporated into the production lines. Gallium arsenide cells are as high as 32% efficient, but their cost still limits them to high-value space applications. Even the thin-film cells have reached efficiencies over 16%, with more improvements in sight.

Our survey revealed that there are several good solar cell systems to consider for an updated solar power satellite. They vary significantly in their characteristics and range from the sophisticated high-efficiency gallium arsenide cells with concentrators to lightweight, lower efficiency thin-film systems on flexible substrates. Each has its advantages. Concentrated systems dramatically reduce the area of cells required and also the total area of the array. Conventional silicon cells combine potential low-cost production with good efficiency. Thin-film systems promise very light weight and good space radiation resistance. There is always the possibility of new photovoltaic materials being discovered as this fascinating, growing industry attracts brilliant young minds.

As we found in our recent survey there are now many good options, so when the time comes to make a firm choice, the designers will be able to pick and choose the best from several candidates.

Wireless Energy Generators—Already in Mass Production

The wireless energy beam capable of sending a billion watts of power to the earth requires many parts to make it work. One of the most critical parts is the radio frequency generator that converts the electricity produced by the solar cells into high frequency energy. This is the heart of the transmission system. During the Boeing studies in the 1970s klystrons were selected as the preferred system because of their high efficiency. Klystrons are used extensively in high power radars. There are other types of tubes that could be used as well as solid-state devices that are similar to solar cells. The frequency selected for the energy beam is 2,450 megahertz, the same frequency as a microwave oven, which uses magnetrons. The efficiency of magnetrons is not as high as klystrons, but they are being made by the millions. Since the studies of the 70s, Bill Brown, who first proved wireless energy transmission, has discovered a simple way to modify magnetrons to make them work as high frequency amplifiers with increased efficiency and greatly reduced unwanted harmonics. As a result it is now practical to use low-cost, mass-produced magnetrons for the radio frequency generators. This change will ensure predictable low cost of the energy transmitter. Today's production cost for a 1,000 watt magnetron is about $12. A one-gigawatt satellite will use about 2 million of them.

A further step toward lower cost could be made if the efficiency and producability of solid-state devices can be improved. The Japanese used a solid-state transmitter in their successful in-space test of wireless energy transmission made in February of 1993.

Developmental Cost Estimates—Political or Real?

Now it is time to discuss cost estimates. This can be dangerous because cost projections are notoriously inaccurate and depend on a set of assumptions that are seldom followed. In addition, inflation plays terrible tricks, particularly when you are trying to project ten to fifteen years or more into the future. However, it is important to understand the magnitude of the costs, what they include, and how they were developed so you can make your own judgment as to their validity. The estimates I will use are those developed during the DOE/NASA studies because they are the best estimates available. I will discuss how they will be affected by evolving technology and changes in the value of a dollar.

In the aerospace industry the estimators handle the effect of inflation by making their estimates in man-hours and material costs by choosing a specific year on which to base their analysis. They then use an inflation factor to project the costs to some year in the future. Most of the cost estimates made by the aerospace industry for their own programs during the early concept phases use cost estimating relationships (CERs). These take into account past programs of similar complexity and account for all the costs of errors and overruns of past programs that will probably happen again.

At this point you might ask, "If that is the case, why is there so much in the press about overruns on government contracts in the aerospace business?" In order to answer that question, it is important to understand how government contracts are awarded.

It is a long and involved process that starts with a perceived need that is either seen by a government organization, such as the Air Force, or an idea that is developed by a contractor to solve a need they think exists. In either case it is not long until a product concept is developed to potentially satisfy that need. Cost estimates are developed based on cost estimating relationships. Invariably the cost is considered too high by the government agency, so they tell the contractor to find ways to reduce the cost. By this time other contractors are aware of the potential product either by being informed by the government or through clever marketing investigation.

The next step in the process is now underway. It may involve contracted studies or company-funded studies. In either event the various contractors involved set about to reduce the cost. The first steps are usually legitimate. They attempt to improve the design to reduce cost. This can be done by reducing the size or by simplification of the design. Generally this is not enough to satisfy the government agency if it is a large program that must be sold to Congress as a line item in the budget. Government procuring agencies know that the lower the projected cost, the easier it is to sell as it moves up the bureaucratic leadership ladder on its way to Congress, so they send the contractors back to the drawing boards to find a lower cost number.

Now the government procurement game starts in earnest and what follows is a sad discourse on our government agencies, members of Congress, and participating contractors. There is nothing about what happens from this point on that can be considered legitimate or honorable. It is only "legal" because it is nearly impossible to prove conspiracy in what happens. The key players are very skilled at the game, while most of the minor participants are unaware of what is going on in the process.

Contractors reduce their program cost estimates in order to satisfy the demands of the procuring agencies even though by this time the contractors have already incorporated all the legitimate costs reductions. So now they must lie. Program managers can hardly tell their employees to lie about the estimated costs, for someone would certainly blow the whistle. So the easiest way to work around that problem is to establish a set of estimating ground rules that will result in a reduced number by not telling the whole truth. It is important to keep the number just low enough to sell the program, because it will be necessary to reduce it some more later, as you will see.

After the studies have defined the program and sold the government on its merit, it is time for the request for proposal (RFP) or request for quotation (RFQ) issued by the government procuring agency, which specifies the requirements for the item they want to buy. Included is all of the criteria that must be met by a successful bidder and how the proposal will be evaluated by the government to select a winner. In most cases, cost is at the top of the list of selection criteria. The contractors know from experience that they must submit the lowest bid to win, or at least low enough to make the final round in the selection process, where they will be given one more chance to make a best and final offer. They also know that the procuring agency needs a low bid so they can convince Congress to fund the program. They know that after a program is funded by Congress it is nearly always continued even when there are serious overruns. This sets the stage for what actually happens.

Both the procuring agency and the bidding contractors are playing the procurement game for high stakes. The objective of the game is to convince Congress to fund the program. The primary rule is to not reveal the truth about actual costs because the program could not be sold to Congress and the contractor would not win the competition. There is an unstated understanding between the government procuring agency and the contractors that they will be able to make up the difference between the proposal costs and real costs through change orders or renegotiation sometime after the contract award.

The next rule of the game is to make the bids look real enough not to raise any suspicion and, most important to the company managers, to stay out of trouble during audit. That is quite easy for those who know how to play the game of hiding real costs from the public, Congress, and auditors. Experienced people don't use real CERs when it comes time to bid on government contracts. They would never win a competitive procurement by using real cost estimates, so most companies bid the cost for which they could do the contract if they did not make any mistakes. The estimate is often made by the bidding team following guidelines from management that tell them to include only the amount of time and materials absolutely essential to accomplish the job. This kind of estimate does not show the real impact on program costs due to error correction. Management then "scrubs" this already low estimate to an even lower level they hope will be enough to win. Their speech to the employees says, "I know it will be tough, but we can do it with hard work," knowing full well that there is not a chance of doing it for the bid price. The important thing is to win the contract; dealing with the overrun will be next year's problem, and the government will pay anyway, so why worry.

With this approach it is quite easy to make the bid one half to one third of what it will actually cost and still be able to defend it in an audit. This works because government procurement agencies are as anxious to hide the real costs as the contractors. Does this sound like the old shell game? It is, and guess who loses.

Unfortunately, there are very few development contracts performed without error, and the inevitable result is large overruns, for which the government (that is, the American taxpayer) pays. Since most development contracts are for the cost plus a fee it is not much of a risk for the contractor. The key risk is in the size of the fee as they are usually given incentives for various items including cost. However, this is usually circumvented by change orders that pad the contract size, so instead of losing fees, the fees are actually increased. This game of how to win government contracts actually results in programs costing much more than if the true costs were stated in the beginning. The combination of initial underfunding and attempts to cut corners results in an increase in the number of errors above what would normally happen if real costs were identified in the contract bids.

Another influencing factor of contractor selection is the government's desire to keep many companies in business in order to preserve a broad technology base in case of emergencies. This generally means that the government often awards contracts to the poorer performing contractors who cannot compete effectively in the commercial marketplace. This in turn often leads to even higher cost overruns.

The last major factor is political influence. Congressmen have now joined the game, and they play with their own set of rules. Since they control the purse strings, particularly committee chairmen, they are in a position to make sure that contracts are awarded to companies within their constituencies or to major political contributors. The procuring government agency knows about this rule and can prepare the way so that the desired contractor can win. "How can that happen in the bidding process?" you ask. Very simple. Write the RFP or RFQ with requirements biased toward a particular contractor. The loser here again is the public.

A prime example of this convoluted contracting process is the Space Shuttle. A senator from Utah was head of the Space Science Committee when the design concept was changed so that the Shuttle would use solid rocket boosters that could be manufactured by a contractor in Utah. This was not the best design, and seven dead astronauts are a testament to that fact.

This is not a new game. It is played all of the time. A classic example was the Air Force TFX procurement program many years ago. There were repeated rounds of competitions with changing requirements until finally they could select General Dynamics, a company with manufacturing facilities in Texas, a state with very powerful political leaders. The result? The F-111, which had many years of major problems, huge overruns, and in the end never really fulfilled the original requirements.

The situation is much different when a firm must compete in the commercial marketplace. There, everything has a fixed price, with an inflation escalator clause in some cases. Performance is guaranteed. If the company makes a large error in estimating costs or performance, they have to absorb the costs; if they cannot, they go into bankruptcy. There is normally no bailout from the government for huge cost overruns in the commercial world. An exception was the bailout by the government of Lockheed over the L-1011 losses and also the losses on the C-5A, which were incurred when the government attempted to make the purchase of that airplane closer to a commercial contract by forcing Lockheed to adhere to the original contract conditions. This bailout was made to keep Lockheed in business because they had so many critical military contracts.

Considering that the development of the first solar power satellite would possibly be a government-sponsored program, why would it be any different from other government programs? It wouldn't be unless it was handled differently from a typical government procurement. Several steps could be taken to minimize the probability of it falling into the same warped pattern as normal government procurement. First, the objective would have to be clearly established and adequate funds committed so there would be no necessity to hide actual costs. This was done successfully on the Saturn/Apollo program. Second, the procurement process should be placed in the hands of a new government agency that would be established for this project only and would not yet have developed a deeply ingrained pattern of bureaucratic inertia in its people and policies. Third, fixed-price contracting should be used wherever possible, even during the development phase, with fixed-price follow-on production contracts being the carrot for the contractors to use honest estimates. Development costs should not be the primary evaluation criteria; instead, performance, procurement costs, maintenance costs, and operational costs should be the primary evaluation criteria.

In any event, when all the perturbations caused by government contracting practices are eliminated, the use of cost estimating relationships based on comparable technology of past programs and accounted for in man-hours are quite accurate for projecting the magnitude of large program costs, as long as the estimator knows the complexity of the program.

One thing that is unique to the estimates made for the cost of solar power satellite program development is the fact that it includes *all* costs necessary to bring the system into being and produce it as if everything had to be done for only that program. Let me explain what I mean. The cost estimates include the cost for assembly bases in space; the development of a new transportation system; the construction of a launch facility; the building of factories to manufacture solar cells, klystrons, and other high production components; and all the elements of the satellite and receiver. This is due to the fact that NASA and DOE had taken the approach on the solar power satellite program that all costs must be accounted for no matter how they are financed. This is not done for any other energy system. For example, the cost of building a new transportation system for coal, whether it be a slurry pipeline or new railroad, is never included as part of the cost of using coal as an energy source. Synthetic fuel costs were often stated as marginal cost, which means that it is assumed you do not have to count the capital cost of the facility and equipment to process the fuel.

The development costs that were estimated by the Boeing study team for NASA are similar to those developed by the other team, headed by Rockwell International. They are stated in 1979 dollars, which means that the labor and material costs are what were experienced in 1979. Included are research costs, costs of a major demonstration unit, the first full-sized satellite, and all the supporting facilities. The total is $116 billion. If the cost of the infrastructure that can be used for other applications is taken out, the remaining cost is $33 billion.

The development costs I have given are based on the ground rules and technology base of 1979. Today, even though inflation would indicate the cost should be doubled, in reality it will be dramatically reduced because much of the infrastructure development cost included in the original estimate has been developed for other reasons. The US Space Station and the terrestrial solar cell industry are but two examples. Huge savings will be achieved with the use of a ground-based prototype of the satellite power generation and energy transmission systems. As I will discuss in a later chapter this will save many tens of billions of dollars and will make it possible for the utility industry to fund the development of the power plant part of the system.

While we were studying solar power satellites, President Carter proposed spending $88 billion on synthetic fuel even though it was going to be more expensive than oil. However, with the breakdown of the oil cartel, this program was later abandoned. Eight billion was spent to build the 799-mile-long Alaska pipeline from the North Slope of Alaska to Valdez. That is $10 million per mile just to carry oil across one state. From 1983 to 1993, the US oil companies *by themselves* spent $192 billion on drilling and exploration for oil and gas—with a result of 30% dry holes.

When we consider how much money has been spent on other energy systems without significant benefits, the cost of investing in our future by developing a solar power satellite system does not seem to be so expensive, even if it costs as much as the original $116 billion estimate.

Production Costs are Reasonable

The results of the production cost estimates for a single 5,000-megawatt satellite was $12 billion. This cost was developed in 1979 dollars and includes the cost of the satellite, the ground receiver, and the cost of transporting it to space. This would be approximately $24 billion in 1995 dollars.

Solar Power Satellite

Production Cost for 5,000-Megawatt-Output Satellite

	$Billions (1979)	Percent of Total	$Billions (1995)
Satellite	$4.5	37.5%	$9.0
Space Transportation	3.7	30.8	7.4
Rectenna	3.8	31.6	7.6
Total	$12.0	100.0%	$24.0

The production costs of much of the satellite were determined by estimating the cost of manufacturing relatively few uniquely different parts in massive quantities. The solar cells that are assembled into panels and make up the largest part of the satellite are one key component. The supporting structure is huge, but very lightweight because of the weightlessness of space. It can be built from aluminum, using automated beam builders or assembled from mass-produced beam components assembled by automated machines. Another major satellite element is the antenna parts, which would also be mass produced and assembled into subarrays. The same is true of the receiving antenna on the ground.

One third of the total cost for the system is space transportation, so it is a major factor. Without achieving low-cost space transportation the entire concept is not practical. Space Shuttle is out of the question as the launch vehicle because of its high operational cost and limited payload. It was designed with a huge external propellant tank that is thrown away on each flight, the cost of the solid rocket fuel is exorbitant, and the shuttle orbiter itself is primarily a research tool and not oriented to routine commercial operation. It requires a standing army of support personnel to maintain it, plan its flights, load and unload the payload bay, check it out for its next flight, fuel it, launch it, and monitor its flight.

The cost of space transportation for the satellite system was based on a very large, fully reusable two-stage launch vehicle that was defined to sufficient detail to verify that it can be operated at the required cost levels. Cost estimates for the entire system were based on a solid definition of the hardware.

I have stated that "production costs are reasonable." That does not mean they are cheap. Based on how the costs were estimated and the comparative checks that can be made, the estimates are a reasonable projection of what can be expected when corrected for inflation. The spread of costs among several different categories of hardware and cost elements means that an error in one area will not catastrophically affect the total. In addition, there appears to be as much chance of the cost being less than projected as there is of it being higher.

One of the key findings of the industry survey we made in 1994 was the need to reduce the size of the satellites from 5,000 megawatts to no larger than 2,000 megawatts. The reason is the inability of the electric utility grids to readily handle a single power plant with so much output. If power is unexpectedly lost for any reason the grids must be able the absorb the sudden loss of power. Experience has shown that about 2,000 megawatts is the maximum they can tolerate without having the whole network drop off line. Therefore, it looks like a 1,000 megawatt output is a good size for an updated design.

The cost will be reduced in proportion to the size decrease. In addition, as technologies advance it will be possible to increase the efficiency of the system and thus reduce its size even more. This will provide further reductions in the cost.

Cost was one of the primary issues in 1980 when the government program was suspended. It is still an issue and will remain an issue until the first satellite has been built and demonstrates that the estimates are correct. Testing on the ground in small scale can remove most uncertainties at a very modest cost, but it will take construction of the real hardware in space to be the final proof.

12

Features of the Satellite System

Let me take you on a tour of a solar power satellite. We'll look at one as defined for the Department of Energy in tha late 1970s. It was designed to provide 5,000 megawatts of electrical power to the earth, which is equal to five typical nuclear power plants. As we make the tour I will discuss how technology has evolved over the intervening years and how the new models will change in the future. One change will likely be a reduction in size to 1,000 megawatts capacity to better match utility needs.

As we approach the satellite we see a huge rectangular array of solar cells stretching into the distance, bathed in dazzling sunlight. A shining jewel in the blackness of space. Its frame is hidden in the shadow of the solar array, and at one end is a giant flat disk, textured with millions of small rectangular slots to focus the energy streaming toward the earth 22,300 miles below.

Size of the Satellite

Our first impression of the satellite would be its gigantic size. A few years ago while on a business trip to New York, I was suddenly struck with the magnitude of size. On a brilliantly clear day I sat in a reception room on the fiftieth floor of one of New York's modern office buildings. I looked out over Manhattan Island and realized that a single satellite would cover half of the sea of buildings before me. It was a sobering vision that dramatically impressed me with how large a solar power satellite would actually be—a

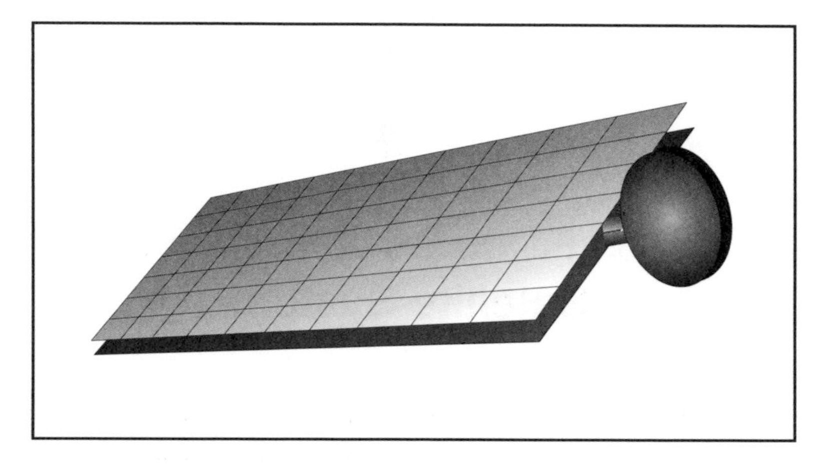

rectangle of nearly twenty square miles (fifty square kilometers), paved entirely with solar cells.

Size alone should not intimidate us since throughout history civilizations have built grand edifices. When I stood in the hot sun of the desert looking up at the pyramids of Egypt, I tried to imagine how such monstrous structures were built by people using only muscle power and rudimentary tools. As I walked the Great Wall of China I had to marvel at the effort to build this barrier along a frontier that wanders for 1,250 miles across the mountain tops of northern China. Built before the birth of Christ, this colossus crosses a mountain 5,000 feet high. Placed in the most inaccessible of regions and built before the existence of engines, it is wide enough to allow five horsemen to ride abreast on its top. The size and quantity of stones used in the construction represent a massive task even by today's standards.

The American interstate highway system stretches for tens of thousands of miles in endless ribbons of concrete and asphalt. Much of it is beautifully landscaped, sweeping through scenic countryside and rugged mountains. I am astounded as I realize this great system was built in only two decades.

Size is really just numbers of smaller things added one at a time. In Manhattan, the size reflects thousands of buildings side by side, each built of millions of parts. The Great Wall of China is tens of billions of rocks laid in neat rows. In the case of solar power satellites, the size will result from billions of solar cells, each not much thicker than a sheet of paper, connected one to the other and hung on a frame in the weightlessness of space.

Even though the satellite is vast, it will not cast a shadow on the earth. Most of the year, the tilt of the earth's axis causes the satellite to pass either above or below the earth's shadow. During this same time, as the satellite's orbit takes it toward the sun from the earth, its shadow would also pass above or below the earth. During the equinox period, when the earth's axis is perpendicular to the sun, the satellite will pass between the earth and the sun. However, even then its shadow won't pass over the earth because of an interesting phenomenon of light. As an object is moved away from the earth toward the very large sphere of the sun, the sunlight passing on one side of the object will converge with the light passing on the other and there will be no shadow.

Also, despite the satellite's size, it won't be visible from earth during the day. The question of visibility depends upon several factors including distance, size, and reflectivity. I have already discussed its size and altitude, so the remaining question is how much light will it reflect. During the daylight hours, the satellite is oriented toward the sun, away from the observer, while the dark side is facing the earth, just like the new moon. We cannot see the new moon during the day and will not be able to see the satellite either.

After dark, the story will be different. With the satellite on the opposite side of the earth from the sun, it will reflect light back toward the earth like the moon. However, since the surface is covered with solar cells that absorb light while converting sunlight to electricity, there will probably be just enough reflected light to make the satellite as visible as an average star. On a clear night, we should be able to see a starry string of pearls—with each satellite representing one pearl—glowing in the night sky.

When I speak to the public about solar power satellites, one of the questions I can always expect is this: "Skylab fell back to the earth; since the solar power satellite is so big, won't it fall too?"

Skylab was America's first large space station and was placed in a low-earth orbit. Low-earth orbit extends from about 75 miles altitude up to 400 miles. Manned systems in low-earth orbit are placed within a maximum of 400 miles because of the radiation in the Van Allen Belt above that altitude. The initial orbit of Skylab when it was launched in 1973 was about 270 miles above the earth, over 22,000 miles closer to earth's gravity and atmosphere than the solar power satellite will be.

We think of the atmosphere as having a finite end, but it does not. It simply becomes less and less dense until finally it is gone. There was a very thin layer of atmosphere in the Skylab orbit, which produced some drag and gradually reduced its speed with atmospheric friction. With the speed decrease, altitude was lost. As Skylab drifted into lower altitudes the atmosphere became denser and eventually friction caused the satellite to burn as it reentered the dense atmosphere. If there had been enough fuel on board to make course corrections, it would still be in orbit.

In the case of a solar power satellite in geosynchronous orbit, the atmosphere is nonexistent so the satellite will not be subjected to atmospheric drag. Satellites at that altitude will stay there for hundreds of thousands of years. There is no danger of a satellite falling from geosynchronous orbit.

The Satellite Framework

As we move behind the satellite and look underneath the vast expanse of solar cells we see a spidery framework of triangular-truss beams that form the satellite's skeleton. It's a rectangular framework, divided into 500-meter-square bays, providing the foundation for mounting all of the other elements and defining the satellite's shape and size. Triangular trusses have proven to be the lightest, most efficient skeleton for very large structures. The great rigid dirigibles built in the early part of the century used aluminum triangular trusses for their structure, and most radio and television antenna towers are truss beams. Such beams are ideally suited for the structure of the satellite, which must be very large and rigid, but will be lightly loaded.

An alternative structural approach we have been looking at for the new smaller models (1,000 megawatts) would use a tetrahedron space frame made of tapered aluminum tubes. A tetrahedron looks like a pyramid with a strut at each corner and around the bottom. All of the struts are the same length, and when you join these pyramids together by adding struts to the tops it becomes a space frame. You may have seen this type of structure used in modern buildings where the structure is left exposed because of its unique beauty. The Museum of Flight in Seattle is a glass-covered space-frame structure that is able to support full-sized airplanes suspended from the structure.

Space structures do not have to carry the crushing loads imposed by gravity; neither do they have to resist the wind or carry huge burdens of snow. In the benign environment of space there is no oxygen or rain to cause corrosion. Size bears very little correlation to weight. A graphic example of the difference between earth-based and space-based structures can be made using the Space Needle, a well-known landmark built in Seattle for the 1962 World's Fair. The needle is a 550-foot tower with a revolving restaurant on top. Built of thousands of tons of steel and concrete, it took a crew of fifty men six months to assemble. A 550-foot section of the main structural frame of the satellite, while about the same physical size as the Space Needle, would weigh less than two tons and take about an hour for one man to assemble using an automated assembling machine.

There are several choices of materials that could be used for the structure, but the most likely will be aluminum. It is lightweight, easy to work, inexpensive, and has the great advantage of long life. The satellite structure would be designed to last indefinitely. Only the power generator and transmitter would require regular maintenance, probably on an annual basis by robotic means supported by maintenance personnel living on a nearby space station.

Advanced composite materials might reduce the weight and increase the rigidity of the satellite, but there are still some unanswered questions about their longevity in the space environment. NASA has been testing a number of new composite materials in space for several years, but until sufficient time has passed to determine life capabilities, aluminum remains the best choice.

Solar Cell Power

If the structure of the satellite is its skeleton, then solar cells are the muscles that do the work. When we look at the cells closely we see smooth, flat, blue-black wafers with a fine grid pattern on their surface to collect the electricity that is generated when sunlight releases electrons within the cell. They convert 16.5% of the sunlight to electricity and are only two thousandths of an inch thick, which is no thicker than a sheet of paper.

In 1978, my Boeing study team faced quite a challenge trying to decide which solar cells to use in our research. At that time there were several different types being developed, but the most common were single crystal silicon cells and gallium arsenide cells. After many sessions with some talented aerospace engineers, the decision was made to use single crystal silicon cells because of their proven performance, light weight, and demonstrated efficiency. The engineers at Rockwell International, who were also studying solar cell choices at the same time, chose gallium arsenide for their studies. Their decision was to explore the potential of more advanced, higher efficiency cells. Our selection was more conservative—we knew the cells were readily available and would work.

As we look at the satellite you can see that the solar cells are assembled into conveniently sized panels, one meter square. What you can't see is how the cells in each panel are interconnected using 14 cells in parallel with each parallel group connected in series to the next group. This provides very high reliability since any four cells in parallel can be lost before the panel stops operating. If a meteorite struck the satellite, it would only experience power loss in the damaged cell area. Based on NASA's meteorite density data, the study team projected a loss of less than 1% of the cells over a 30-year period. The new models will have even higher reliability as they will use the technique developed by the terrestrial solar cell industry of bypass diodes to provide for an alternate electrical path. Each panel would be interconnected like the squares of a quilt to fill each structural bay. These in turn are joined together to create the rectangular form of the satellite.

Great strides have been made in solar cell development since the 1970s, and there are now many good cell materials from which to choose. Single crystal silicon is still the most common, multilayer gallium arsenide are still the highest efficiency, but thin-film cells made from cadmium telluride, copper-indium-deselenide, or multilayer amorphous silicon are lighter weight and less expensive, but also less efficient. The decision for the future must be made whether to build a large satellite with many low-cost cells or a smaller satellite using fewer, but more expensive cells. Another way to reduce the number of cells required is by using concentrators, such as mirrors or lenses, to focus the sunlight from a large area to a smaller area of solar cells. Regardless of the type of solar cell material selected, the cells would be installed in a way similar to the efficient arrangement worked out in the 1970s studies—a smooth and shiny quilt in the darkness of space.

Keeping the Satellite in Place

By now you may have noticed that the only moving parts on the entire satellite are the rotary joint between the solar array and the transmitting antenna and the attitude control system, which keeps the satellite pointing toward the sun. Because the solar cells must always point toward the sun while the transmitting antenna points toward the earth, the connecting joint will make one revolution each day. At the same time, this joint must also provide for the transmission of eight billion watts of electrical output from the solar cells to the transmitter. Unfortunately, wires cannot be used as they would soon twist off due to the continuous rotating motion. Slip-rings, similar to those used in some electric motors and radar sets, but on a much larger scale, would work well here. Since the motion is slow, there should be very little wear. As a comparison, the engine of an automobile will make more revolutions during a 15-minute drive to the grocery store than the antenna will make in a hundred years.

The task of keeping the satellite pointing toward the sun will be performed by an attitude control system. The attitude control systems for the Space Shuttle and for some of the current satellite systems burn small amounts of chemical propellants in tiny rockets to turn and position the spacecraft to the desired position. However, on the solar power satellites we can take advantage of electricity to power ion thrusters to accelerate an inert gas (most likely argon) to very high velocity, thus creating a rocket-thrust reaction.

The concept of using electricity to power rockets was demonstrated many years ago when Hughes Aircraft Company developed ion thrusters for NASA. These rockets used electricity generated by solar cells. In addition to the development of ion thrusters, work has progressed on other types of electric propulsion systems.

The attitude control system would also be used to maintain the satellite's precise position in geosynchronous orbit. Due to slight gravitational variations on the earth, one of the anomalies of geosynchronous orbit is that a satellite left without any control will wander from one location to another within the orbit. It is necessary to keep the satellite aligned with its earth receiving antenna while simultaneously ensuring that it will not interfere with weather and communications satellites already in geosynchronous orbit.

Argon, used by the attitude control system, would be the only material on the satellite that would have to be resupplied on a periodic schedule. Current geosynchronous satellites must carry all additional fuel needed for their useful life when they are launched since there is presently no capability to resupply them. The infrastructure necessary to resupply argon and carry out routine maintenance on the satellites would be an integral part of the concept development.

The Transmitting Antenna

As we turn our attention to the transmitting antenna we see that it is mounted on a rotary joint at one end of the satellite and faces the earth at all times. It seems small compared to the rest of the satellite, although a disk one kilometer (0.62 miles) in diameter is large by most standards. Closer inspection reveals the disk to be covered with aluminum planks. As we take a still closer look we see that the planks are covered with slots, which are actually slotted wave guides. Wave guides are hollow rectangular box sections that channel radio-frequency energy so it can radiate out through the slots in the front face of the guides. This design concept is identical to many large phased-array radars in use today. The difference would be in the specific frequency used and in the fact that radar sets transmit energy in pulses, receiving reflected signals between pulses. By contrast, energy from the satellite would be continuously transmitted. The term "phased array" means that the beam formation and steering comes from control of the radio-frequency waves across the face of the transmitter.

As we make our way to the backside of the antenna we see it is cluttered with supporting structure and with electronic equipment generating radio-frequency energy. The transmitter is the most complex part of the satellite system with two important functions to perform. It must convert electric energy generated by the solar cells to radio-frequency energy and it must form the beam.

Conversion of electricity to radio waves can be done in several different ways. One example is a device called a magnetron, which is similar to a vacuum tube and is used in today's microwave ovens. Another means of conversion is called a klystron and is used most commonly in large radars with power outputs up to a million watts. Newly developed solid-state devices similar in concept to transistors are also a possibility and are being developed by the Japanese for their 10-megawatt, low-earth orbit demonstration satellite.

The satellite we are looking at uses high-power klystrons for the energy conversion, but new models will use much smaller magnetrons or solid state converters. The decision of which type of conversion method to use will be based on conversion efficiency and life expectancy. The goal is to achieve more than 80% conversion efficiency. The higher the efficiency, the smaller the satellite can be. A life expectancy of 40 years is typical for power generating plants, but in the benign environment of space, the satellites can possibly last 100 years or more.

Transmission of the energy to earth is the last function the satellite needs to accomplish to fulfill its role as a solar power plant. Control of the beam is accomplished by controlling the frequency phasing of the radio waves over the face of the antenna. This in turn requires controlling the phasing of each individual microwave generator in relation to its neighbor.

Satellite Vulnerability

From our brief tour, you can see the elegant simplicity of the design, but before leaving the subject of the satellite itself, I want to address one often-raised question about the vulnerability of the solar satellite. Even though the cold war is just a memory and the threat of nuclear holocaust has diminished with the collapse of the communist world, the question still remains: "Isn't the solar power satellite vulnerable to attack?"

In this case a solar power satellite can be compared to a commercial ship, only instead of traveling on the great oceans of earth, it will move through the high seas of space. Throughout history, the right of commercial ships to ply the oceans of the world in relative safety has been respected by other nations. An attack on the high seas, even of a commercial vessel, has always been considered a deliberate act of war. This same logic should and will be preserved as we move into the new frontier of space. As on the high seas, the rights of sovereign nations and their vessels in space will be respected and an attack would be considered an act of war.

Throughout the world, there are power plants situated in well-known locations. Some, like Grand Coulee Dam and Hoover Dam, are even major tourist attractions. None of these power plants are defended against attack. Even though most have some kind of fence against intruders, the best precautions will not keep out dedicated terrorists, as evidenced by antinuclear protesters who illegally enter nuclear test sites. Our power plants are neither more nor less vulnerable to terrorist attack than any other public place, as we have seen in the devastating World Trade Center explosion in New York City in February of 1993 and the bombing of the Federal Building in Oklahoma City in April of 1995. Power plants are as vulnerable to enemy attack by air or ballistic missiles as any potential target in our nation. The best deterrent is having a large number of power plants. No nation would have the combined weaponry and skill to bypass our defense systems to destroy them all.

The location of the solar power satellites is a strong deterrent to attack. Access to geosynchronous orbit is difficult to achieve, and even with modern rockets it is a journey of over five hours. High-powered laser weapons could conceivably be used in the future, but they require high technology and a large sustained power output that is difficult to achieve.

Vandals or terrorists would also find it very difficult to attack a solar power satellite. Even the receiving antenna on earth would prove to be a frustrating target because of its great size.

If the United States were the only nation to possess solar power satellites, we could be in a situation envied by other nations and in that case it might be prudent to consider defending them from attack. However, the best defense is to make energy from solar power satellites available to all who need it. The energy crisis is not only an American crisis but a world crisis as well. Global access to energy from space would benefit all nations and preclude a threat of war.

13

Features of the Energy Beam

The miracle of the solar power satellite energy system is built around the concept of transmitting huge amounts of energy over thousands of miles without the use of wires. Wireless power transmission has been the dream of many people, but today technology is making it happen. The energy beam has no moving parts, it cannot be seen, it will pass effortlessly through the atmosphere and clouds, it will be very large to keep the energy density low, it will be safe, it will be environmentally clean, and it will be an efficient transmitter of energy.

As you saw on our tour of the satellite the transmitter is a flat phased-array antenna. The radio-frequency energy is distributed over the face of the antenna in slotted wave guides that allow the energy to be radiated into space in a tightly controlled beam. Even though it is mechanically pointed toward the earth, the final steering of the beam is accomplished electronically by controlling the phasing of the radio waves across the face of the antenna to guide it to the receiver on the earth. The actual controlling signals would come from the receiver site on the ground to keep the beam always pointing precisely at the receiving rectenna on the earth.

Safety of the Energy Beam Comes First

There are several important considerations to take into account when designing a wireless power transmission system for solar power satellites. The very first is safety. In order to ensure a safe environment on the earth the energy density is limited to a level that would be safe for all life forms. In the original studies an additional energy density limit was established to prevent excessive heating of the ionosphere. The maximum limit as initially established was 23 milliwatts per square centimeter, which is 230 watts per square meter. This level is one fourth of the energy in bright sunlight on the earth.

The next consideration is the selection of the best radio frequency to use. The most important factors are to select a frequency that will have minimum reaction with the atmosphere and one that will not interfere with wireless communication systems. Radio frequency bands are used for many purposes. They include broadcast radio, television, marine radios, commercial radios such as those used by taxicabs, police radios, military radios, radars, energy systems such as household microwave ovens, medical diathermy machines, industrial dryers, telephone systems such as communication satellites and cellular telephones, and amateur radio. Frequency allocation is made by the federal government in accordance with international agreements with other countries. Without control there would be chaos with users interfering with each other.

The frequency selected for the wireless energy transmitter is in the Industrial Scientific & Medical (ISM) band. The specific frequency selected was 2,450 megahertz, which is the same frequency that is used in microwave ovens and medical diathermy equipment. The atmosphere is nearly transparent at this frequency. An added advantage is the broad experience base developed through building and using huge quantities of radio frequency energy generators at this frequency. Another frequency that could be used is 5,800 megahertz, but it is subject to more atmospheric interference.

After selecting the frequency and the maximum energy density in the center of the beam as it reaches the earth, the engineers are then able to determine how big the transmitting antenna and the receiving rectenna must be. Radio-frequency beams follow fundamental laws of physics, with the key controlling factors being the distance the beam travels, its frequency, and the size of the transmitting antenna. These factors define the beam size and are independent of the how much energy is transmitted. The amount of energy transmitted then determines the magnitude of the energy density.

With geosynchronous orbit as the satellite location and 2,450 megahertz as the transmitter frequency, that leaves only one variable, the size of the transmitting antenna, to determine the size of the beam when it reaches the earth. If we were to reduce the size of the transmitting antenna, the beam would become larger on the earth and the energy density would be reduced. Conversely, in order to make it smaller and more concentrated on the earth the transmitter must be larger on the satellite.

After selecting the transmitter diameter, the antenna designer's job is to tailor the profile of the energy density in the beam from the center to the outer edge. At the same time they would decrease the side-lobe energy. The side-lobes contain the energy that forms as concentric rings outside of the main beam. They are at a very low level, and the better the antenna design the less energy they contain.

The diameter of the satellite transmitter for a 5,000 megawatt output satellite was established at one kilometer (0.62 miles) in diameter in order to maintain a maximum of 230 watts per square meter of energy density at the center of the beam. This very low level was selected for all the studies to provide consistency of results. However, later tests by the government using the huge radio telescope at Arecibo in Puerto Rico demonstrated that excessive heating of the ionosphere would not occur, so that limit can be relaxed.

During the late 1970s a large number of environmental studies were conducted under the auspices of the Environmental Protection Agency to determine the effects of radio-frequency energy on life forms. After the termination of the satellite studies in 1980 many of these studies continued since there are so many uses of radio frequency energy. The US Air Force was particularly interested because of the great number of high-power radio and radar systems they operate. As a result they have been the chief sponsors of the recent studies. I had the opportunity to attend the First Annual Wireless Power Transmission Conference (WPT '93) in San Antonio, Texas, where an entire session was devoted to reports on the findings of these studies. They covered experiments at the biological cell level to determine if there were any carcinogenic effects that would cause mutation and cancer. They included investigation of higher life forms up to and including apes.

The findings of all the studies have been consistent. The only adverse effects found are due to heating when the energy level is high enough to overcome the ability of a cell or organism to reject the excess heat. This level is not reached until the energy density is in the neighborhood of 1,000 watts per square meter or higher. This far exceeds the maximum level of 230 watts per square meter established for all of the solar power satellite studies. The threshold is different for different life forms because of how they absorb the energy and how well they can reject the heat. Past experience with high power radars has shown the most sensitive parts of the human body are the eyes and liver. If we are exposed to high levels for extended periods of time, we could develop cataracts, or in extreme cases the liver could be damaged. There have been a few cases where people exposed to very high-power radar transmitters with energy densities much higher than the level in the energy beam have suffered such effects.

Experiments made at the cell level showed there was absolutely no evidence of carcinogenic effects at the energy levels considered for the wireless energy beam. These experiments were made by scientists who were very familiar with the known effects of ionizing radiation from x-rays and nuclear radiation and from chemical agent exposure. In addition, testing conducted on honey bees and birds have also shown no permanent effects after exposure to energy levels about four times greater than the 230 watts per square meter study limit and minimal effects during exposure to these levels.

The evidence is very good that the wireless energy beam will be totally safe. It is also likely that a level of 500 watts per square meter, about twice the study level, could be used safely. Increasing the allowable energy density to 500 watts per square meter would have definite economic and environmental benefits by reducing the size of the rectenna.

If someone by some chance found themselves on top of the antenna right in the middle of one of these beams, they would not suffer any damage. Bill Brown, the inventor of the concept, has been working with wireless power transmission research since he demonstrated his first working system in 1964. He is now retired from Raytheon but is continuing his work to further develop the system. He has a working demonstration model that he uses to show wireless power transmission in operation. His transmitter is about thirty inches square and has an energy density higher than our original study level of 230 watts per square meter. When he gives his demonstration he has a rectenna with lights that come on and change their brightness in response to the energy pattern they receive from the transmitter on the opposite side of the room. As he gives his talk he casually walks through the beam and the lights go out as his body absorbs the energy. He will stand there for a while as he talks and then as he moves out of the beam the lights come back on. It is a very graphic demonstration that he has done for many years, and he is an elderly gentleman. He has commented that he can just feel a slight warming effect.

So far we have examined the safety of the beam if something went wrong and people were exposed to the maximum level of radio frequency energy in the center of the beam. However, it is very unlikely that this could happen since the beam will always be pointing at the rectenna. The real question to address is the normal everyday environment near the rectenna site.

At the edge of the antenna the maximum energy level will be one watt per square meter or one tenth of the US standard of acceptable exposure to radio-frequency energy. To prevent casual exposure even at this low level, an exclusion fence will be placed outside of this area at the point where the energy level drops below one tenth of a watt per square meter or one hundredth of the US exposure standard. This would be the maximum level people would normally ever be exposed to. Even the side lobes of the beam, which form as concentric rings of ever-decreasing levels of energy around the beam, will not exceed this level.

The Rectenna—The Biggest Opportunity

The satellite (the power generating part of the system) is located in space, but the element that will be of greatest interest to us on the ground will be the receiving antenna. Since it is very large, we must ensure that it can be designed, built, and operated in a manner acceptable to all of us.

Let's take a look at what the rectenna (receiving antenna) would look like for a satellite like the one we toured. The receiver will convert the radio-frequency beam from the satellite into 5,000 megawatts of electricity. If we want to keep the energy density low and still receive that much power, the beam has to be spread out over a large area. The actual size was determined when we established the maximum energy density at 230 watts per square meter and applied all of the natural laws that apply to radio-wave beams. Using this criteria we ended up with a beam 5.9 miles in diameter when it reaches the earth. This is a variable that can change as the system evolves in the final design phase—in fact, we are currently thinking of reducing the output to 1,000 megawatts, which reduces the beam diameter to two miles.

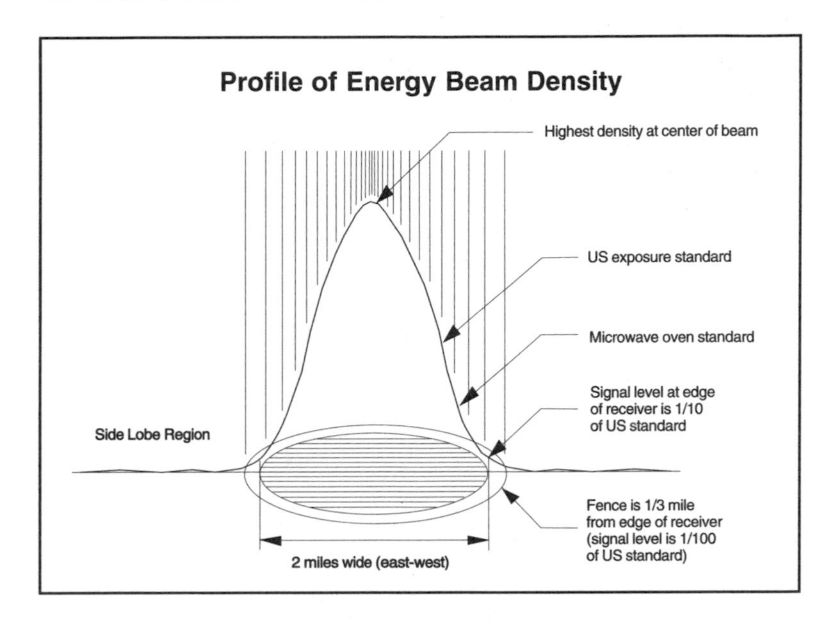

The profile of the energy in the beam when it reaches the earth would be shaped like a bell. In the middle of the beam the energy is at its maximum. As we move from the center the density falls off rapidly, and by the time we reach the perimeter the density is down to one milliwatt per square centimeter. This is the minimum density that would be economical to recover. For a 5,000-megawatt satellite, the exclusion fence would be another 0.8 miles from the edge of the rectenna; for a 1,000-megawatt satellite, the exclusion fence would be one-third of a mile from the edge of the rectenna.

Because the beam comes from geosynchronous orbit around the equator, it intercepts the earth at an angle in the United States. The angle of the beam would be the same as the degrees of latitude of the receiving site. Because of the angle the rectenna must be oval in shape with the long dimensions running north and south.

In its simplest form, the rectenna would be made up of a wire-mesh back-screen, mounted perpendicular to the beam with antenna elements mounted at regular intervals in front. There are many ways that these elements could be designed, but typical would be small half-wave dipoles mounted a few inches apart with a rectifying diode for each element. The dipoles receive the radio-frequency energy, and the rectifying diodes convert the radio energy directly to electricity.

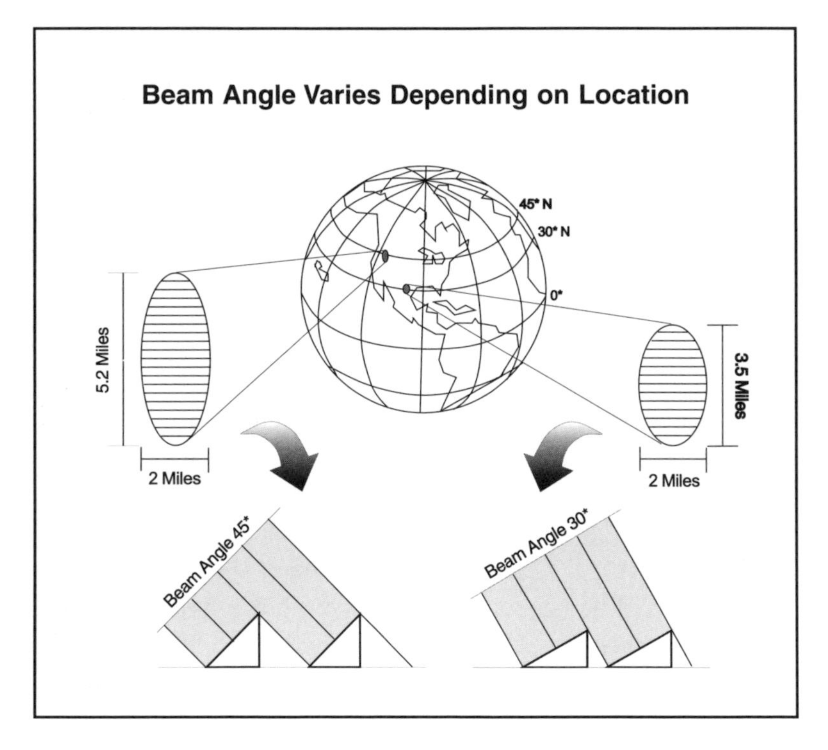

All of this would be supported on columns and beams above the ground. Because of the angle of the energy beam to the earth, the rectenna can be built in rows with each row of antenna elements inclined towards the satellite, leaving space between each row. It will intercept more than 99% of the microwave energy and convert it directly to electricity by use of diodes, while allowing most of the sunlight to pass through. As a result, it would be possible to use the land underneath the antenna for agriculture.

An ideal location for a large rectenna would be in the desert near Las Vegas, where we find only a few lonely roads and some coyotes, very few people, and nearly level land of very little use today. About the only objections we might find would be from the military for cluttering up their lonely test areas, or from a few old prospectors still looking for the end of the rainbow.

Another site might be further north in a cold, dry area in the west, possibly near Cheyenne, Wyoming. At this site there are rolling hills and grazing land for cattle; hay land along river beds, a few ranches, some back-country roads, and a trail or two. Even though there may not be many inhabitants, they are undoubtedly fiercely independent and will need some powerful persuasion to be willing to leave their land.

These remote sites will disturb few people, but it is going to be difficult to find areas large enough for the receiver sites close to population centers where there is the largest demand for power.

If we are going to be part of the problem-solving team that is developing the system, let's see how we might approach this obstacle. First let us consider the problems of finding a site that covers about 30 square miles (for 5,000 megawatts) or 6 square miles (for 1,000 megawatts) near a big city where we would like to locate a receiving antenna. Population densities make this a problem. More than likely, when we found a potential area in the northeast it would be covered with farms, a stream or two, some scrub land, several patches of trees, and a few roads. If this is the case, the people who live there will be pretty upset if we suggest moving them from the farms where they grew up and spent their lives just so that New York City can have bright lights. But what if we could offer them a great opportunity for the future and possibly a new place to live and work without having to move?

Even though the amount of land required for the receiving antennas is large, it only takes one fourth the land required by coal strip-mining for an equivalent amount of power. Site selection would be a compromise between the advantages of a remote location and the distance required for the transmission lines to bring electricity to the populated centers. Further development of superconductors would make it possible to place all the rectennas in remote areas since the length of the transmission lines would no longer be a problem. However, with a little change in the design of the receiver, we could have an exciting new concept that would build for the future in more ways than just providing energy.

John J. Olson, an industrial engineer, artist, and designer, let his imagination look at the problem in a very creative way. His resulting idea is fascinating.

What would happen if there were a series of greenhouses with glass roofs made up of wire screen and antenna elements imbedded in heavy glass, permanently tilted toward a satellite in geosynchronous orbit? With a simple modification to the design, the rectenna could become rows of greenhouses covering the entire site of the rectenna. Since the energy in the radio-frequency beam would be converted to electricity with only the sunlight passing through the antenna, we could safely use the area under the glass roof for agriculture. We could install tracks to handle automated farm machinery and build in the capability to collect rain water. Tempered glass would protect the antenna from damage from snow, hail, and the occasional bird that might think a half-wave dipole antenna element was a fat, juicy worm.

With the rectenna modified to become a vast series of greenhouses we will be able to grow ten times as much food as is possible in open farming, using only one tenth as much water—enough food to feed more than a million people. Some crops, such as wheat, do not grow much better in a controlled environment, but most do much better. For example, some crops like cucumbers are phenomenal; they produce about fifty times more in a controlled environment than they do in the open. We could flood the world with pickles! Even "crops" that are not presently practical can be grown in a controlled environment. Did you know that shrimp can be raised that way? Not in the earth, of course, but in controlled environment tanks where their growth rate is spectacular. It would even be possible to grow plants to produce liquid fuel with biomass conversion.

With the rectenna built as greenhouses, instead of taking valuable land out of production, we will be multiplying its production potential. The possibilities are endless. Think of the opportunity to satisfy many of the needs of a populated area with one facility. Providing not only electrical energy, it could also supply most of the food and part of the fuel for a city of several million people. The long-range implications for the world are staggering. It could someday be possible to feed starving people at a level that would mean freedom through abundance. And if, in desert areas adjacent to an ocean, you add a desalinization plant that uses the abundant electricity to produce large quantities of potable water, great fields of greenhouses could turn an arid area such as the Baja Peninsula into a garden spot.

You might ask, "If this is so good, why don't we build huge greenhouses now?" We do in a few places, but the capital cost of a greenhouse is very high. By the time a farmer borrows the money, builds the greenhouse, buys his equipment, and then pays the mortgage, he just about breaks even. But if we need to build the structure anyway for a receiving antenna, the additional costs of a greenhouse become quite reasonable.

Even the Displaced Will Prosper

Let's go back and consider the problem of procuring the site near a big eastern city. With the greenhouse option, we can now offer the displaced farmers a chance to keep farming their land, but with greatly enhanced potential. With plentiful low-cost energy and reasonably priced food, the benefits could far outweigh the disadvantages even to those directly affected.

How about those sites out west? That cold weather around Cheyenne could be tamed when we combine the attributes of solar electricity generated in space with direct solar heating on the ground in our greenhouses. The cattle might become very lazy basking in the warmth of their glass-roofed homes eating to their heart's content while the snow lies three feet deep outside. Even the most independent-minded of ranchers could appreciate the advantages of raising cattle in that environment. Some of the finest tasting beef in the world feeds on the warm luxuriant grass of coconut palm plantations in the tropics of Vanuatu, in the South Pacific.

Meanwhile, down near Las Vegas, they would no longer be gambling with nature when green things start growing out there on the desert in glass houses. Since the amount of water necessary to grow plants in a greenhouse is reduced to 10% of what it takes in the open, even the meager amount available in the desert could be enough to grow crops in places never before possible.

Another exciting possibility would be to grow plants for biomass conversion to liquid or gaseous fuels. Since plants use carbon dioxide to grow, would it be possible that they could use it at the same rate they would later be releasing it in the burning process? Sounds a little bit like perpetual motion. It might even be possible to raise fast-growing plants to provide raw material for the manufacture of paper products and reduce our dependence on rapidly disappearing forests.

Another option for minimizing the impact of land use for the rectenna sites is to enlarge the satellite transmitter and allow the maximum energy density to be increased to 500 watts per square meter. If this were done, the area of the rectenna could be reduced to about 15 square miles for 5,000 megawatts and 3 square miles for 1,000 megawatts. Then we need to challenge the antenna designers to tailor the antenna characteristic to make the profile of the energy density more like a barrel rather than a bell, which would make the energy density more uniform over the area of the rectenna without increasing the maximum level. The result would be another decrease in area required, maybe as small as 10 to 12 square miles for 5,000 megawatts or 2 to 2.5 square miles for 1,000 megawatts. At these sizes the number of available sites would increase dramatically, so that rectennas could be located nearly anywhere.

14

A Development Plan

How do we go about creating the fourth era of energy—at the moment merely a vision? How do we accomplish such an enormous task? What is the first step? The Department of Energy, the government agency responsible for developing new energy systems, abandoned the idea in 1980 and has been unwilling to reevaluate the concept since. Their position was made clear in a letter I received in February of 1995 from a Deputy Assistant Secretary of Energy in response to a letter I sent to the Secretary of Energy. I had urged the Department of Energy to reevaluate the solar power satellite concept in light of the major advances that have been made over the years since 1980 and to establish a program office to coordinate with other interested government and commercial organizations. The DOE's letter stated:

> For over two decades, the Department of Energy (DOE) and its predecessor agencies have supported various technologies that could increase domestic contributions toward our nation's energy needs. Among the energy concepts investigated and terminated were ocean thermal systems, wave energy systems, ocean

> and river current systems, solar ponds, solar heated wind towers and, as you point out, space solar systems.
>
> All of the foregoing systems are technically feasible. However, as with most research investments, choices are necessary. Typically, options have been eliminated because energy costs were judged to be high, energy contributions were likely to be limited to only a few regions, reliability appeared to be low, estimated development costs were high, or other risks to commercial success were projected.
>
> In line with the President's program to reduce federal spending, the Department is developing plans to reduce program costs by more that $10 billion over the next five years. Given these commitments, opportunities for initiating new programs are limited and we are generally not able to offer encouragement for federal funding for space power. Of course, where reasons are compelling, we will strive to accommodate. However, for the most part, it will be necessary to complete or terminate programs where possible and to maximize returns from investments in continuing programs.

It is apparent that the Department of Energy is in a business-as-usual mode that does not include any effort to develop new energy sources.

On the other hand no single commercial company has sufficient resources to undertake the task, and as yet, no industrial leader has emerged to draw together a partnership of companies to do the job. Yet we cannot ignore the urgency as the cost to the environment and to our economy continues to escalate and devastate.

A recent correspondence I received from Dr. Bruce Middleton, former head of Australia's space program and now managing director of Asia Pacific Aerospace Consultants Pty Ltd, put the potential impact in perspective. He wrote:

> The Intergovernmental Panel on Climate Change has predicted that the likely increase in average temperature by the period 2030 to 2050 will be around 2.5 degrees. For the US alone, the economic impact of such a change is projected to amount to

> $60 billion plus per year. This cost cannot be avoided simply by reducing carbon dioxide emissions, for it is estimated that sequestering the carbon dioxide produced from fossil fuel combustion for power generation in the US would increase the cost of electricity by 30% to 100%. At 50%, this would add $85 billion per year to US energy costs. . . .
>
> The environmentally preferred solution to this problem must be to replace coal-burning power generation capacity with orbiting arrays of solar cells delivering energy to the earth's surface by radio frequency transmission.

The sooner we face the need for a new clean energy source, the less we will lose in money spent for ever-escalating fossil fuel costs and repair to our damaged environment. If enough people raise the issue of developing a new energy source, the politicians will be forced to pay attention. If enough people become aware there is the potential for an abundant, low-cost, clean energy source, the debate for its development will start. Public awareness will raise the issues to be explored, resulting in a free discussion of ideas, for and against.

When people become aware of the full potential of solar energy from space and what it could mean in their lives, they will demand of our leaders that they proceed. With a mandate from the people, the leadership can step forward with confidence knowing that the public is ready to support the effort. Then we can take the first steps!

There are two primary paths that can be followed to develop solar power satellites. One is a government program and the other is commercial development with some government support. In 1980 the only conceivable option was a massive government-sponsored and funded program. Today that is no longer the case. Advances in the enabling technologies along with significant infrastructure development now makes possible commercial development of the program with some government support.

There are legitimate arguments for each approach, but the many pitfalls along the path of government development makes commercial development much more desirable. However, for the sake of argument I will describe both options along with some of the advantages and pitfalls.

The Government Development Option

A logical approach for a concept this large is for the federal government to fund the development as a national resource. This has been done often in the past and is one of the normal functions of government. However, government development of solar power satellites will only be successful if the President of the United States supports it, as did President Kennedy in sending men to the moon and President Eisenhower when he proposed the interstate highway system. For this alternative to work it must be a *total commitment*, made without hollow political gesture or reservation. It is an enormous and challenging task that will not succeed with half-hearted efforts. It will require long-term support with sufficient funds to carry the program through the inevitable hard times that will occur as development progresses.

The basic problem with this alternative is the fact that a solar power satellite program is a commercial energy system. People are reluctant to have the government involved in commercial ventures, as historically they nearly always become a mismanaged fiasco. Two exceptions of successful national programs are the interstate highways and the moon landing, which were truly national programs that could only be done by government. However, there are good reasons to look to government for the solar power satellite program because of its size, the fact that it needs development of multi-use high-technology infrastructure, and more importantly because of its broad international implications.

Choosing the people and organization within the government to lead such an effort becomes the critical factor in order to avoid two nearly fatal traps. One trap is assigning the responsibility to an existing agency that has developed a deep bureaucratic mindset through long routine government service, and the second would be to choose leaders who do not understand the commercial utility market.

To avoid these traps, the best solution is to establish a new agency under the leadership of an individual selected from industry with the specific responsibility for this project only, reporting to the executive branch at a very high level. The agency needs to have authority to perform all necessary functions to accomplish the goal, without other distracting responsibilities. The task is to develop an operational energy system. It is not a scientific research endeavor, but is a massive engineering effort, based on highly scientific principles to generate commercial power at low cost.

In this development option the primary function of the government is to manage the program, select industrial contractors to perform the necessary work, coordinate international agreements, and to supply the funding. A very critical part of this function is planning the program, establishing specifications for the various program elements, and selecting the contractors to perform the tasks. By far the majority of the actual work will be accomplished by contractors. Most of the problems with government-run programs start with the specifications for the tasks and selection criteria for the contractors.

Through the years government bureaucracy has developed the policy of writing specifications for government procurements that limit the flexibility of a contractor to deliver the best product because the specifications define a solution rather than a need. They also apply more specifications than required to protect the bureaucrats from any possible blame if something goes wrong. In a commercial system the purchaser specifies what he wants his product to accomplish and allows the manufacturer flexibility in how to best achieve that. The satellite program must operate in a commercially competitive environment and therefore has to be designed from the beginning for commercial use. The requirements must emphasize system performance, ease of maintenance, and low operational costs. How best to achieve the results should be left to the ingenuity and expertise of the winning contractor for each system element.

A gross comparison of the requirements specified for a commercial airliner as opposed to a government-procured airplane is likely to progress along the following lines. The commercial customer might specify: "The aircraft shall have the ability to carry XX number of passengers at YY minimum seat spacing, over a minimum distance of ZZ miles, with the fuel consumption no greater than AA pounds per mile, and the airplane shall be warranted for 60,000 flight hours and 20,000 landings. It shall be certified flight-worthy according to appropriate government requirements and delivered on or before such-and-such a date."

The government specification is likely to say: "The aircraft shall have a wingspan of 150 feet and a maximum gross weight of 400,000 pounds, carry a payload of 100,000 pounds, and have a cargo bay 12 feet wide, 40 feet long, and 8 feet high. It shall use company XYZ engines, number 1234. It shall be built with material Unobtainium-543 and coated with Invisibilium, Specification 1000.987-123D. It shall conform to all military specifications listed in the following references and their subreferences: AA, BB, CC, DD, EE, FF, GG, HH, JJ, KK, and LL. The contractor shall report all costs down through four WBS levels. Time-keeping records shall be maintained according to procedure XA12.52 and recorded to the nearest tenth of an hour. . . ."

That is only the beginning. The government will specify the development schedule, the test program, the number of maintenance people to plan for, and anything else the specification writers can think of—I imagine you get the picture. One would think their jobs depend on the volume of paperwork generated instead of a balance sheet to measure their success or failure.

Let me give you a real example by comparing the development of the C-5A and the 747. The 747 initial design was developed by

Boeing in response to an Air Force competition for a large military transport to be called the C-5A. Boeing lost the competition to Lockheed, who signed a contract for the C-5A in August of 1965. Instead of totally discarding the work they had done in preparation for the competition, Boeing decided to modify the design from a military airplane to a commercial transport of the same size. Boeing presented the preliminary design to the airlines and on April 1, 1966, received a letter of commitment from Pan Am that justified the start of commercial development. Boeing developed the 747 under commercial requirements in three years and eight months from start of development to the inaugural passenger service flight, while Lockheed developed the C-5A under restrictive government requirements that took four years and four months before the first airplane was delivered to the Air Force. The 747 flies faster, carries more cargo weight, has nearly twice the range, and costs less to buy and operate than the C-5A.

The story of the Space Shuttle is an example of how wrong a government procurement can go. The original concept was to develop a fully reusable two-stage vehicle. Each stage would be able to fly back to its launch base. The system was to be designed for minimum maintenance and rapid turnaround to achieve low per-flight cost for an operational system.

Unfortunately, the government made all the classic mistakes during its development cycle. First of all it was supposed to be an operational system to provide low-cost space access, but was developed by NASA, a research and development agency with no commercial experience. The managers placed in charge were mainly professional bureaucrats or technologists, while many of the experienced leaders of the Saturn/Apollo program had retired or returned to industry. There was serious intercenter rivalry as the various NASA centers worked to change the configuration to favor their center. Instead of using proven low-cost elements to achieve an effective operational system the technologists saw the opportunity to develop new high-technology components. The politicians holding the purse strings saw it as a huge pork barrel and shaped the design to favor contractors in their areas. Annual funding was limited (a typical practice in government procurement) and forced design decisions based on compromise. This in turn limited the development of low-cost operational systems and favored initially cheap systems with future high operational cost. The experienced bureaucrats made sure they could not be blamed for anything.

In addition, the detail specifications that evolved were oriented to using the system as a research device rather than an operational system. The final configuration was a hybrid design that incorporated the worst of all these factors. Today we have a Space Shuttle that works some of the time, in spite of everything. But we went through the tragic trauma of watching the Challenger explode, carrying her crew to their deaths. Then the gut-wrenching investigations to determine cause and blame. Finally a fix that is really only a series of splints patching things back together. The Shuttle is two orders of magnitude more expensive to operate than it should be and will never be able to meet its original operational goals.

The development of the solar power satellite system cannot be successful if it is developed in the same environment as the Space Shuttle.

With planning and specifications completed, contractors can then be selected to perform specific tasks. For a system this large and complex it will be essential to have an overall systems integration contractor to supplement the government agency management. It must be a contractor that not only has the ability to integrate large systems, but more importantly, with the experience and understanding of commercial power plant operations. There is no company in existence today with all of the necessary skills, so it will be necessary for the successful bidder to assemble a team of contractors, capable of working together, that will bring all the needed skills to the task.

Integration of a large complex program is not handled well by a government agency. Just look at the disastrous cost and schedule overruns experienced by the Washington Public Power Supply System (WPPSS) in their effort to manage the construction of a group of nuclear power plants in the state of Washington several years ago. The end result was the abandonment of the program after completion of one out of the five funded plants. WPPSS then defaulted on the billions of dollars of bonds sold to finance the rest of the project. Even the Saturn/Apollo government integration team ran into trouble. After the Apollo 4 fire the director of NASA recognized that there had been insufficient integration and control of the overall program and turned to an industrial contractor to oversee testing, integration, and evaluation to bring the program to a successful completion. Development of the Space Station was in serious trouble under the direct management of NASA, and it was not until NASA consolidated management under a prime contractor that the cost and schedule was brought under control.

The inclusion of other nations of the world as partners in the development of solar power satellites has many potentially desirable benefits. Government interaction will be necessary for international agreements controlling satellite slot assignments in geosynchronous orbit as well as traffic control in space and radiofrequency agreements. Sharing of developmental costs would be another asset. Weighed against these benefits is the problem inherent in spreading responsibility into areas that are difficult to control. It is a serious question that needs much consideration before making a final decision. The one clearly effective way to involve international cooperation is to open the competition for design, development, and manufacture of the various elements to companies from all nations. This would provide distribution of the development without losing control.

The schedule for completion of the development program, including assembly of the first demonstration unit, is another critical decision. It is important that sufficient time be provided to accomplish the development and thorough testing of all components prior to committing them to production. However, the schedule must also be as short as possible in order to maintain a sense of urgency among all participants. The contracts should be heavily incentivized to maintain delivery schedules, using both penalties and bonuses. There are so many individual elements to fit together that a delay of any single one can seriously impact the cost and schedule of the entire program. A period of 10 years is a reasonable time schedule for development. Once a schedule is established there should be no thought of later change. It must become the one inviolate milestone.

In the government development option all the funding for the design, development, manufacture, and testing of the satellite system and all the supporting infrastructure would be funded by the government and would culminate in a full-size, operational satellite.

The funding could and should be financed by a tax on imported foreign oil and/or a tax on other current power generation systems that are causing air pollution in our atmosphere. This has two benefits. It taxes products that are used by energy consumers who will ultimately receive the benefits, and it makes the price advantage even greater for a nonpolluting energy system, thereby accelerating its development and implementation.

This option, though viable, is fraught with potential pitfalls. It requires extraordinary leadership and commitment to be successful. Similar problems face the option for commercial development, but the very nature of the competitive commercial marketplace is self-controlling. A poorly managed company soon falls by the wayside and another takes its place so that the whole operation is not damaged. This does not happen in government procurement where poor performers are allowed to continue on the team, thus impeding the whole operation.

Commercial Development Option with Government Support

During the late 1970s as I worked to have the solar power satellite program move from studies to full-scale development, my only choice was to push for a massive government program of the type I just described. As you know it was an unsuccessful effort, but much has changed through the ensuing years. I retired from Boeing in 1987 to pursue the adventure of sailing the oceans of the world with my wife in our sailboat, setting aside the dream of bringing solar energy from space to serve the people of the earth. After several years wandering through the South Pacific we stopped to spend the hurricane season in the safety of a sheltered mooring in Brisbane, Australia. While we were in Sydney visiting friends a feature article in the *Sydney Herald* about international interest in solar power satellites showed me that the dream was not dead, only waiting for someone to bring it into reality.

We abandoned our plans to sail on around the world so that I could once again pick up the challenge of developing solar power satellites. I started writing this book, and in the summer of 1993 I also formed a company whose purpose is to pursue the commercial development of solar power satellites.

I assembled a team of experts, and the company's first project was to determine the status of the required technology and plan how to develop the system as a commercial venture. It was an exciting time meeting with many of the people involved in the early studies and those who are working in related fields today. We made a survey of the photovoltaic industry, national laboratories, and representative utilities. From the resulting information we put together a plan for commercial development of solar power satellites. It will require government support in some critical areas, but the only government funding required is some seed money and some multi-use technology development. The necessary funds can be found within currently planned government expenditures by focusing their efforts on solving the problems associated with solar power satellites rather than general research.

The plan described in the following pages is the one our company developed and is presently working on. It is based on an industry/government partnership with industry taking the leading role to develop the power plants. The important role for government will be to coordinate international agreements, support the development of high-technology multi-use infrastructure, and assume the risk of buying the first operational satellite.

Only government can establish international agreements on orbit slot assignments, frequency allocations, space debris cleanup, space traffic control, and licensing. And there is still the question of whether commercial investors will be willing to finance the development of a new reusable space transportation system for solar power satellites prior to proving the system is economical. It is also still desirable to have the government assume some of the development risk on the first unit and to be the focal point for international cooperation during the development phase, but most of the financing and control can be commercial.

Shown below is a 10-year schedule for commercial development of the satellite system.

The primary focus of the early part of the program is to develop and validate the system on the ground with a small-scale engineering prototype. The ground test program brings together the solar cell technology currently being developed for terrestrial photovoltaics with the evolving technology of wireless power transmission.

The approach of using a ground-based prototype to do the major development testing has resulted in a dramatic reduction in the projected development cost and is one of the key elements making commercial development possible. The program consists of a small-scale terrestrial-based solar cell array (in the range of 50 to 250 kilowatts peak output) coupled to a phased-array wireless power transmitter, which would transmit the energy over a short distance (one to five kilometers) to a receiving antenna (rectenna), then feed the DC power output through an inverter/power controller into a commercial AC utility power grid.

Each element of the system will be designed to incorporate several different technology approaches to be tested in the complete end-to-end test installation. The installation will duplicate all aspects of the power generating systems for the solar power satellite concept except for the space environment and the range and size of the energy beam.

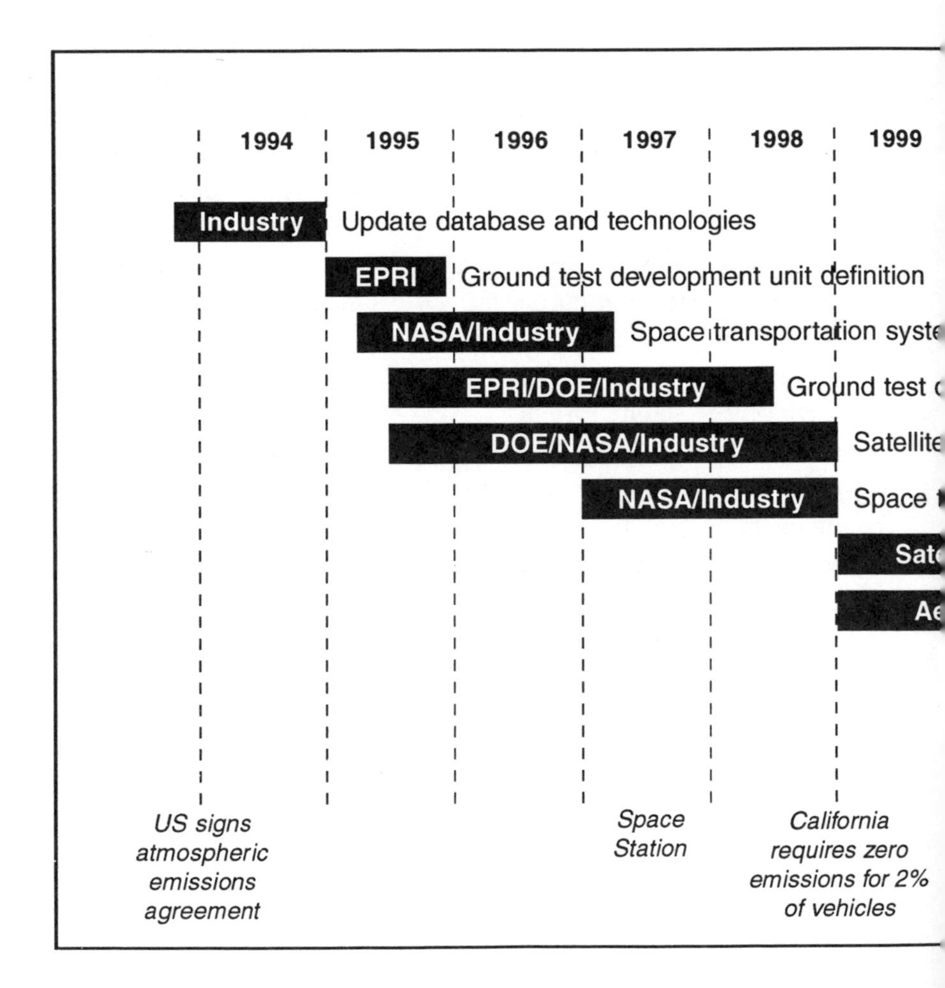

Testing for the space-oriented aspects of the concept is a logical mission for Space Station. The Space Station is a major piece of the infrastructure needed to develop solar power satellites and is currently being developed as a national investment. By focusing the research conducted on the Space Station to solve the problems of developing the space aspects of solar power satellites, NASA would still be able to accomplish their space research objectives with very little increase in cost. Most of the space research needed for the solar power satellites is also needed for any other space

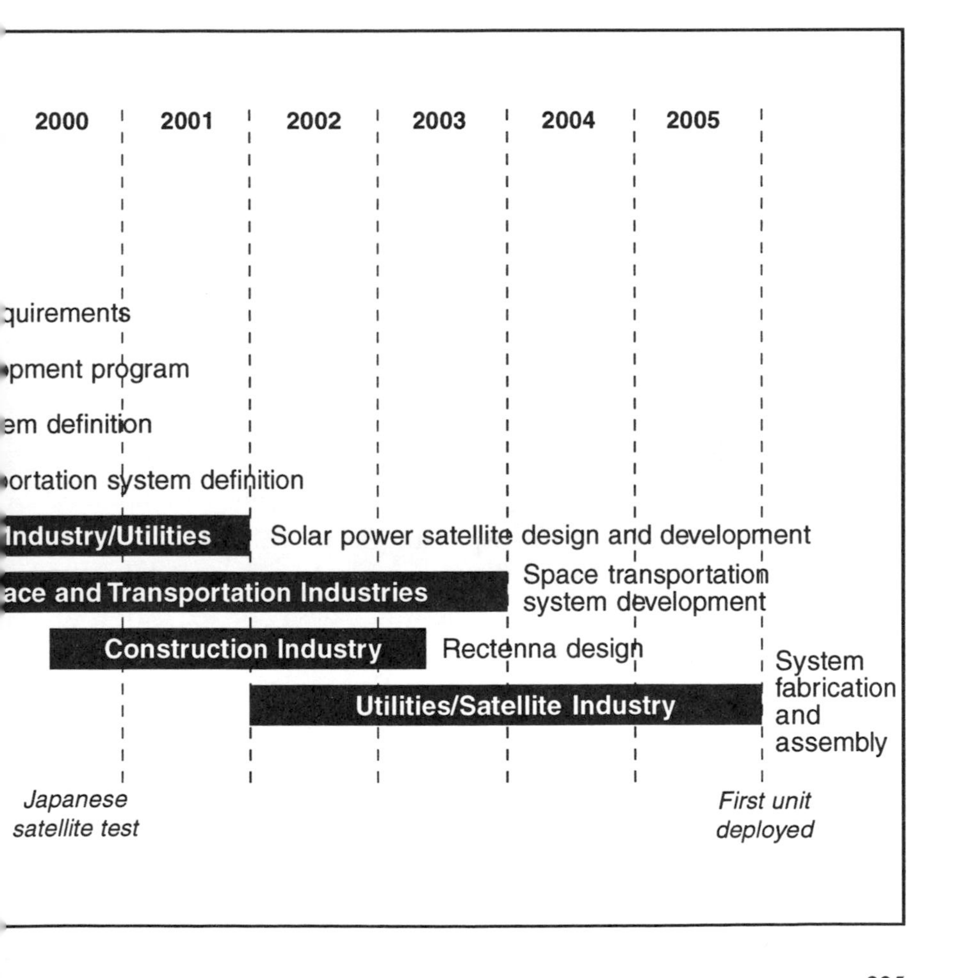

program. The economic benefits of using the Space Station for developing technology for solar power satellites will give it a clear mission that will more than justify the cost of its development.

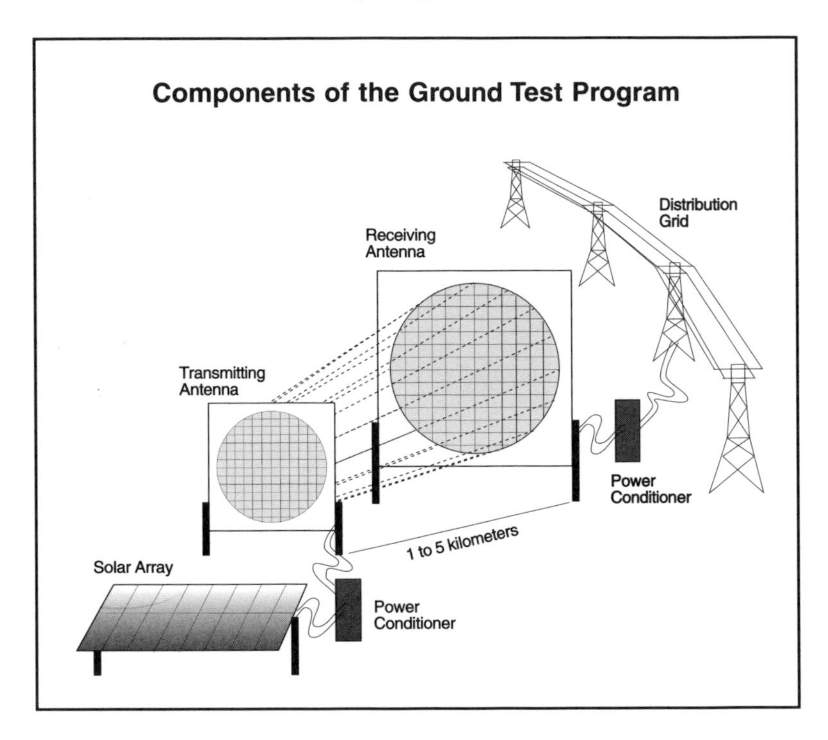

The most expensive part of the program will be the development of a new reusable space transportation system. The need for a low-cost space transport is not unique to the solar power satellite program. NASA is currently working with industry on the early phases of a program to demonstrate a small-scale prototype of a new low-cost reusable system to replace the Space Shuttle. There are several space programs planned that would benefit from a new low-cost launch vehicle. One example is Teledesic's plan to launch 840 satellites for telephone communication. However, none of the planned programs are large enough to justify the cost of a new system. What is unique about the solar power satellite program is that it is large enough to justify the development of a new low-cost system by itself. The potential space transportation market is huge. For example, if solar power satellites were only used to replace worn-out power plants the annual revenue for transporting them to space would be over $15 billion per year. This only addresses the US replacement market. If the total world market is considered, the space transportation revenues would be closer to $100 billion a year. It is certainly a large enough market to entice competitive commercial operations. To be successful, however, it is very important that the requirements for transporting the solar power satellite hardware be incorporated into its development.

The final part of the development plan is a full-size operational solar power satellite that proves the validity of all facets of the concept, including the most important—cost.

The question is, who pays for all of this? Funding can come from many sources, but the following is what I see happening.

The basic premise is an industry/government partnership. In this scenario the government establishes program offices in its major affected agencies: Department of Energy, NASA, Department of Commerce, Environmental Protection Agency, State Department, Department of Defense, and the Federal Communications Commission. These offices are formed to act as focal points within their area of responsibility and to coordinate international participation where applicable. Their funding responsibilities would be limited primarily to providing seed money for program planning and definition, multi-use technology development, conducting environmental impact testing, and funding space testing.

The primary source of funding for the ground test program should be supplied by the utility companies, either directly or through the Electric Power Research Institute (EPRI). However, the Department of Energy should also provide some of the funding with seed money to initiate the program. Industrial hardware manufacturers who will benefit from the enormous market being developed should also contribute. I estimate the total cost of this part of the program would be in the neighborhood of $50 million dollars over a three-year period.

Establishing the requirements for a low-cost space transportation system should be funded by NASA but developed by a commercial company outside the aerospace industry in order to avoid institutional bias and bureaucratic bungling. The question of who should fund the cost of developing the space transportation system is the toughest question to answer. By far the best answer is for it to be a purely commercial development. However, the problem lies in the fact that it needs to be developed in parallel with the development of the satellite and before the system has proven to be economical. As a result there is not the guaranteed market that is so essential to entice financial sources to commit the required risk capital—particularly since it is very expensive. There are some mitigating circumstances that may make it possible to obtain commercial financing. First of all, the ground test program will be complete and the confidence level in the solar power satellite system will be quite high for both cost and efficiency, therefore reducing the risk that it will not be cost competitive. Second, there are several other potential market requirements for low-cost space transportation that can justify developing a new system. As a result the overall risk may be acceptable for commercial development.

If commercial development is too much of a risk, there are other alternatives. One is for the government to guarantee paying for a minimum number of flights per year to support military and NASA launches. This would significantly reduce the risk for a developer and still save the government money. Another option is to have the government develop the system as a national resource. The best solution, however, will result from commercial development, rather than government.

In the past, launch vehicles have been developed by the aerospace industry for the government or to launch commercial satellites, but in either case the launch of the vehicles was usually supported by the company that designed and built them. This is not the approach that would be used for launching solar power satellites. The space transportation industry needs to adopt the pattern used by other transportation industries. In all cases the manufacturers of the vehicles, whether they are trucks, ships, airplanes, or railroad locomotives, sell their products to an operating company. The operating company then uses the vehicles to haul cargo or people.

A space transportation company, similar to an air-cargo airline, is the logical purchaser of new reusable space freighters. Financing could be handled in the same way an airline finances the purchase of their airplanes. The aerospace company that developed the spaceliner would face a situation similar to developing a new airplane. They would have to finance the development cost and sell the vehicles for a price that would allow them to recover their investment over a reasonable number of deliveries. The key issue is having a large enough market to recover all expenses and make a profit.

The situation is similar for the satellite, with some variations. Most of the technology development will have been accomplished by the ground test program with space testing supported by NASA on the Space Shuttle and Space Station. Testing on the Shuttle and Space Station should be funded as part of NASA's basic budget. The key to financing the remainder of the satellite development is also the market. In this case the critical step is placing the order for the first operational satellite and a commitment that if it meets its cost and efficiency goals there will be more orders. A government-owned utility such as Bonneville Power Administration is the logical buyer of the first unit. Bonneville, with more than 20,000 megawatts of generating capacity and a large distribution system, is large enough to readily absorb the power from a 1,000 megawatt power plant. In addition, there are sites within their service area where the receiving antenna could be built. The cost will be repaid by the revenue generated by the satellite. The main reason a government utility should buy the first unit is so the government would accept the risk.

The price of the first unit would cover the cost of the satellite and a portion of the design and development cost. The developing contractors would be expected to recover the remainder of the development cost over the sale of some reasonable number of follow-on orders.

In a commercial development scenario there will be a single overall prime contractor who will manage the development of the satellite system. Supporting the prime contractor will be numerous companies from all over the world. It is very likely that many of these companies will be sharing in the financial risk of developing a new large system in a manner similar to the way commercial aircraft manufacturers share the risks with their major subcontractors when they develop a new airplane. The magnitude of the risk will likely be too large for a single company to undertake.

During all of this development cycle there would be major participation by both foreign governments and international companies. Individual hardware development can be done by companies that have experience in the design and development of similar systems. Several aerospace companies exist in the US, Europe, Asia, and Russia with experience building rocket-powered vehicles and airplanes. Some have extensive experience in building numerous large commercial transports as well.

In the case of the satellite hardware there are several worldwide companies that are manufacturing solar cells. The transmitter design is very similar in concept to the large phased-array radars manufactured by many companies.

The ground rectenna requires the experience of large construction companies rather than the aerospace industry. They would have to be familiar with the handling of electrical power and its processing, as well as capable of manufacturing the vast number of individual rectifying antenna elements.

Site surveying and selection for the ground receiver will be a major task requiring great tact. As discussed earlier it will require a good deal of persuasion to displace people who are located on potential sites without creating animosity. The task required for the development phase is to find one site for the initial full-size demonstration unit, but it cannot stop there. Sites must be identified and commitments made for their eventual development to make sure that follow-on units can be built. The most likely location for the initial unit is on one of the remote United States government-owned sites in the desert area of the Southwest or the Pacific Northwest.

One potential site outside of the United States is on the Baja Peninsula of Mexico. A few years ago as we drove from San Diego to Cabo San Lucas, on the tip of the Baja, we passed through a vast desert area. Between villages we came upon a perfect site located near the Pacific Ocean. It was flat as a table, stretching as far into the distance as I could see, with very little vegetation and no visible houses, people, or grazing stock. I envisioned a glass-encased rectenna in the form of row after row of greenhouses. Inside were farmers tending their crops. Energy would be flowing from space to turn this desert into a garden spot. Water to grow the food would come from the nearby Pacific. Ocean water, made pure and sweet by desalination plants using some of the electricity captured by the rectenna. Nearby a community, where today there is nothing, transformed by energy and water, bringing jobs and prosperity to a poor region.

The selection of a launch site for the space freighters will be another interesting challenge for the space transportation company. The site that comes to mind immediately is the Kennedy Space Center at Cape Canaveral. Nearly all of our civilian and commercial launches in the US have occurred there. It could be modified for launching the satellite hardware. However, it may not be the best location.

Since several launches a day would be required to transport all of the satellite hardware to space during the construction period, one of the problems with this site would be the noise. The problem would become worse as the years went by and greater numbers of satellites were built.

The ideal launch location for a space freighter transporting hardware to geosynchronous orbit is on the equator since geosynchronous orbit is directly over the equator. Since Kennedy Space Center in Florida is located at 28 degrees of latitude, any launch from that site starts with an orbit inclination of at least 28 degrees from the equator. It requires extra energy to make the orbit angle change to arrive at geosynchronous orbit, which has no inclination. If the launch is made from near the equator, the total amount of energy required is reduced.

One practical solution to these problems is to establish a launch base on a remote island near the equator. There are several candidates, such as Johnston or Jarvis Islands that are United States possessions in the Pacific Ocean. Another potential candidate is Clipperton, which belongs to France and is located off the west coast of Central America.

My favorite is Canton Atoll in the Phoenix Group of Kiribati in the South Pacific Ocean. It was once a tracking base for the US manned space flights and later a secret Air Force base for tracking ballistic missile tests in the Pacific. It was turned over to Kiribati in 1979 and is uninhabited except for a few people placed there by the Kiribati government as caretakers. It has all the requirements of an ideal launch base and is located within three degrees of the equator, with the nearest inhabited islands about 600 miles away. It is a desert atoll out of the hurricane belt, with consistently good weather and little rain. It has an existing runway in good condition that can easily be extended to a length of 15,000 feet. The protected lagoon has a dredged entrance and a turning basin 35 feet deep with ample room to expand. The atoll itself is large enough to be able to place the housing area a reasonable distance from the launch pads and runway. It could be a virtual paradise for people living and working there, with fabulous fishing and beautiful white sandy beaches.

I spent four months on Canton in 1991, and as I looked out across the lagoon each morning, I imagined I could hear the body-shaking roar of rocket engines coming to life mingled with the high-pitched whine of jet engines bringing the boosters home. There over the end of the runway were the orbiters as they swooped to earth, silent as ghosts, still shimmering from reentry heat. The lagoon was filled with cargo ships awaiting their turn at the loading dock, and lying offshore, moored in the lee of the island, a liquid hydrogen tanker discharging its load of fuel through the underwater line to storage tanks on shore.

Canton has known many pioneering activities in the past as a refueling base for Pan Am transpacific clippers in the early days of air travel, as a large military outpost in the second World War, and as a refueling base for DC-6s and Stratocruisers after the war. It was one of the space-watching outposts that tracked John Glenn's historical orbit around the world. It could once again rise like a phoenix from abandonment to become the center of space-produced energy.

Construction of a remote launch base would add to the transportation cost of materials to the base, but the reduced cost of launching and isolating the noise problem would probably more than make up the difference. I would be first in line to lead the relocation of a work force to this tropical island paradise.

The final goal of the development program is the successful delivery of electric power from the first demonstration satellite into a commercial earth-based energy grid. From this demonstration it will be possible to project the cost of power from the follow-on production units.

In all likelihood, early tests would indicate a successful conclusion with a very high confidence level long before the demonstration unit was completed. The costs of the major elements, like space transportation and solar cells, would also have been proven before completion. It is quite likely that follow-on orders would be placed by utilities, even before the final demonstration.

Market Growth

Normal market pressures will determine the rate of expansion of the system. The first orders will probably be motivated by the need to replace old nuclear and fossil fuel plants that are wearing out. There have been few new electric power generating plants built since the 1973 oil embargo, so nearly all of the existing fossil fuel plants and more than half of the nuclear plants will soon be in need of replacement due to age. In addition, cost of electricity will also become a major influence as inflation raises the cost of fuel to exorbitant levels. As we convert from fossil-fuel–burning systems to electric systems, the demand for electric power will require many more satellites to fill the demand.

One of the changes that will certainly occur is the growth in the electric utility business. This will place a high demand on financial institutions to help finance the growth. The rate of growth will accelerate as the cost of energy from solar power satellites decreases in comparison to other sources.

The foreign sales market could take some interesting twists. With the technology developed and the concept proven, other countries will want to build their own satellites. This would be well within the capability of several nations. One way they could do this is to purchase the necessary support systems from the United States companies, as well as many of the critical components, and do the assembly themselves. It would not be much different for them to buy a reusable space freighter from the US than it is to buy an airplane for their airlines. The rectenna, which represents about 30% of the total cost, could easily be constructed by anyone. They would probably need a license to manufacture the individual antenna elements, or they could purchase them separately. It is also very likely that some countries could manufacture their own solar cells. The number of potential variations is infinite, the degree of involvement being determined by each country's technical and manufacturing ability.

The Fourth Era—The Era of Solar Power

As more satellites are placed in service the United States will experience a sustained period of economic growth. Employment levels directly engaged in the industry will remain high, and the multiplying effect in support businesses will be a significant part of total United States employment.

Infrastructure is one of the catch words of modern society, but it accurately defines the vast array of industries and support systems that must be developed to be able to manufacture and deploy solar power satellites. Space transportation systems that move beyond the Space Shuttle to fully reusable systems capable of being flown daily like an airliner. Habitats in space that can house hundreds of workers. Robotic machines to assemble the satellites in space. Factories designed to manufacture billions of component parts. Construction companies that understand how to cover uneven ground with glass-paned structures manufactured in vast quantities.

The word infrastructure for solar power satellites is really a synonym for *wealth*, because the money spent for energy delivered from solar power satellites goes into the infrastructure that will create it and distribute it. There is no cost for fuel. Payments for the cost of energy will go to the people who build and operate the system. None of it goes to a foreign land for fuel that is pumped out of the ground, burned, and is gone forever. When the satellites start delivering energy from space it will be a stream of wealth pouring from the sky. Wealth that will come to everyone in the form of low-cost energy. Wealth that will enable America to provide a prosperous home for all her people and to overcome the debt she has accumulated. High-technology spin-offs will create other new businesses and elevate industry in the United States to new levels of capability. In a developed nation such as ours high technology is our competitive edge.

At some subtle point in the future, oil will lose its dominant grip on world energy use and its role as the price setter. That will be the dawn of the fourth energy era—energy from space.

Sun power will be the global solution for the coming energy crisis. Atmospheric pollution will be eliminated and once again we will be able to see the horizon in sharp brilliant colors. There will be no more excessive accumulation of carbon dioxide in our atmosphere, and we will cease heating the earth. There will be energy for the developing nations to emerge from the shroud of poverty. We will be able to look forward to a future economy based on abundant energy with no built-in inflation driver. We will not be at the mercy of an unpredictable foreign government; we will have finally achieved *energy independence.*

We will be able to look to the future with confidence as we expand into the high frontier of space.

15

The New Frontier

It is the unique nature of humanity to try and reach beyond ourselves, to strive for knowledge we do not have, and to make our own mark on the world that has led to the incredible expansion of humankind's dominance of this world in a time span that is but a blink of an eye in relationship to the age of creation. No other creatures on earth have these characteristics. By controlling our environment, we humans have made it possible to live nearly anyplace we desire. We do not change our ability to survive in different environments; we simply go beyond ourselves and make the environment suitable to survive. With the subjugation of energy to our will, we have been able to greatly multiply our ability and reach out to ever more distant horizons.

As we look back in history, we find that humanity is always searching for a new frontier to explore and develop. If we do not find one we become restless and try to take one from our neighbor, which often results in war.

And there are those who are not satisfied with the status quo, who have to sail beyond the horizon, climb the highest mountain, seek out the depths of the sea, search out the mystery of the atom or the magic of electrons running around inside a tiny chip of silicon. These are the explorers—whether of geography or of science—who are looking for new frontiers. Some are small and private; others change the course of the world. I believe it is worthwhile to examine some of these examples to see what might be happening to us today.

The Early Explorers

Columbus had a bizarre dream and convinced Queen Isabella that he could reach the Orient and all its great riches by sailing west. After she gave him his ships, he and his reluctant crew started out, much against the wisdom of the day that said they would fall off the edge of the world. Even though he found a new world, Columbus fell into disfavor and never did reap the benefits of his great discovery. Others followed to develop the land and to colonize—some to loot the riches, many to start new lives, others to seek freedom. They were the misfits, the mavericks, the restless, and the builders who developed new nations and reaped rich harvests of benefits undreamed of by Columbus. Eventually, in the colonies, it was the settlers who came and brought stability to the land where they would build their homes, businesses, farms, and eventually the greatest economic giant on earth.

Later, within the new nation called the United States of America, two explorers named Lewis and Clark convinced Congress to fund an expedition to the Pacific Northwest territory. Into unexplored wilderness they went, finding the glories of the Rockies, the Snake River, and the Columbia Gorge, returning with tales of wonder and excitement. In their paths followed the fur trappers, the traders, and rugged visionaries. These men could see beyond the wilderness to the day when settlers, industry, and commerce would thrive and the riches would flow. Flow they have, again far beyond the dreams of those early explorers and developers.

Wilbur and Orville Wright dreamed of flying like birds through the heavens—and they did. Once they proved it could be done, more followed, but in those early days it was difficult to imagine any way that flight could be of any practical use except for the personal thrill of flying. A few visionaries saw the day when the mail could be carried successfully, or even passengers. The military became the first to exploit the potential, however, and the skies of World War I were alive with the angry snarl of Spads and Fokkers amidst the rattle of machine guns. After the war, the airplane once again became a toy, but a much better toy. Gradually some of the early visions began to take on the semblance of reality. A few passengers were being carried and some of the mail was being delivered by air.

Then in 1927 Charles Lindberg electrified the world with his flight across the Atlantic, and aviation had come of age. Shortly thereafter, cabin stewardesses flew for the first time on the elegant three-engine Boeing 80A of the Boeing Air Transportation system (the forerunner of United Air Lines). Soon the Douglas DC-3 revolutionized air passenger service. Some even thought it was the ultimate aircraft, never to be exceeded.

The magnitude of today's commerce in the sky boggles the mind. At any given time, there are probably more than 3,000 commercial passenger and cargo aircraft in the air somewhere in the world. Even if they only average 100 passengers each flight, that represents more than 300,000 people cruising the skies of the world at any time. What amazing progress for a frontier opened less than a century ago to the skepticism of many highly educated people who pronounced it a lot of foolishness and waste.

The lessons to be learned here are typical of the evolutionary development cycle of all new frontiers, whether geographical or technical. This cycle begins with exploration, frequently the result of one person's dream. More often, it is the culmination of the efforts of several dedicated people striving to accomplish that which has never been done before. It is a period of excitement and thrills. Even the uninvolved are intrigued and watch in awe as spectators. They may even contribute in a monetary way. Once the goal has been reached, however, their interest wanes. The unknown is now known. It was fun while it was happening, but what good is it?

The next part of the cycle is much more difficult. This is the development phase. All the excitement of the initial exploration is gone. Very little glory is left to the second person who does anything. Not enough is known about the new frontier to be able to accomplish anything very useful, except to gain further experience and knowledge. It is hard to find anyone anxious to fund an activity whose objective is simply more experience and knowledge.

Now come the visionaries—those people who can see from the scant base of available knowledge into the future of practical application. They clearly see the benefits and know that an investment made in developing experience and knowledge will someday lead to large returns, even though the development period can be decades in length with only limited returns starting immediately. Sometimes military exploitation stimulates the development cycle as it did with the airplane.

As the development cycle progresses and the experience and knowledge base expands, the mists of uncertainty recede. Now there are many who can see the potential benefits. At that time the cycle can enter the next phase, which is the exploitation phase. This is exploitation in the truest sense: *to turn to practical account; utilization for profit*. Exploitation occurs when a new frontier is turned into a practical, solid place to conduct day-to-day commerce. Now it is time for the people to participate and benefit. Opportunity seems to occur at every turn. Benefits materialize from sources not even imagined by the original explorer. The new frontier no longer exists and in its place is an established part of a mature but growing society. It is accepted as if it had always existed. Today hasn't air travel become the most common way to travel long distances?

The length of the cycle varies, but not as much as one might think. In modern times—let us say the last 400 years—the cycle has varied from a minimum of 30 years to a typical length of 50 years from the initial discovery to the start of the exploitation period. This necessary time period has not accelerated at the same rate as technology. The reason for this is not a function of technology capability but rather a function of human characteristics. These characteristics have not changed dramatically throughout the centuries even though the state of accumulated human knowledge has changed. The proportion of dreamers and visionaries to those who are dubious of new ideas remains the same. As a result, the course of progress is strewn with the obstacles of doubt. As the level of technological frontiers continues to rise, the decision-making time remains fairly constant. The length of time required to reach the exploitation phase of the cycle is really dependent upon how clearly and convincingly the visionaries can paint a picture of reward to those who must make the developmental investments.

The New Frontier

Through the years the peoples of the earth have explored every continent, walked the beaches of every island, climbed the highest mountains, and peered into the depths of the sea. The last great frontier for mankind to explore is the heavens above. So where is our new frontier—*space*—in this cycle? The term "space" may be somewhat misleading because even though it appears to be a great void, it contains many things: the earth, our moon, the other planets, asteroids, the sun, and billions and billions of stars in the heavens. This frontier is so enormous we can't even conceive of its vast limits. Let us look at how our generation is reaching across the void to exploit this grand frontier.

The exploration of space began on October 4, 1957, with the launching of Sputnik, and reached its pinnacle on July 20, 1969, with the first manned landing on the moon. That is the recent history most of us remember. It was an extremely difficult achievement that earned the praise and respect of the world. Even though there is continuing exploration, from that moment on, space was in the development phase. In the mind of the public, however, the goal was accomplished and then the questions began. What is the moon good for? What will space do for me? What are we going to get for our investment? Were the moon rocks worth all those billions of dollars?

I knew that the romance of exploration in space was over during a moment of truth in Bastrop, Louisiana, in the fall of 1969. I had given an after-dinner talk to the Lyons Club on the Saturn/Apollo lunar landing and was answering questions when an old farmer stood up in the back of the room. His question went like this: "Now, young man, I think it's fine that we sent men to the moon, but what I want to know is when are you fellows going to figure out how to make a good septic tank?"

Over the ensuing years, exploratory space programs have been very difficult to sell to Congress. Developmental programs have not fared much better. Compromise and funding restrictions for the Space Shuttle forced the selection of a hybrid, partially reusable configuration, that has experienced delays, a tragic accident, and high operational costs.

Funding delays and the resulting cost escalation have caused serious delays in the development and operation of the Space Station, which is so badly needed for our country to gain the experience and knowledge necessary to be able to live and work in space. What happened is what normally occurs during the development phase. The excitement is gone and there is no clear understanding of how additional investment will lead to practical benefits. Only a few dedicated and patient visionaries keep things going by looking beyond to see the benefits of more knowledge and experience. Knowledge and experience are the pathways to commercial development. A profitable commercial venture, to succeed, must be reduced to routine operations; it cannot be a high-risk adventure. The necessary experience and knowledge in space is being slowly acquired.

In some areas of our space endeavor the exploitation phase is underway, putting space to practical use. Most of us watch the Olympic games live from anywhere in the world. The fall of Communism was brought into our living rooms as it happened; the breach of the Berlin Wall occurred before our eyes. War in Kuwait was carried onto our TV screens as smart bombs found their way down ventilation shafts and Scud missiles were shot out of the sky. The instantaneous spread of news and communication around the world is primarily due to the ability of communication satellites to literally blanket the earth with their coverage. If you place an international phone call, it most likely will go via satellite. The deployment of US satellites is a private enterprise business under governmental control with the cost of space launch services paid for by the satellite owners. It is now a big and profitable industry and getting bigger. From this first big commercial exploitation of space all of us reap benefits every day.

In the public service area, the Landsat program provides so much information on our land and crops that we cannot analyze all of the data. Weather satellites give us a continuous view of the goings on in our atmosphere. The military depends on space for earth observation, weather, communications, navigation, and who knows what else.

The achievement in 1991 of worldwide, 24-hour-a-day coverage of the Global Positioning System (GPS) has revolutionized navigation for those who travel the oceans of the world. During the war with Iraq many of these units were rushed into production for the troops serving in Desert Storm. With GPS our troops knew precisely where they were at all times as they moved across the limitless, featureless desert.

The early investment in exploration and development of our new frontier is now starting to return solid dividends. This is just the beginning. The potential of space is as large as space itself. Let us close our eyes to the mundane moment-to-moment problems that continuously surround us and let our minds wander freely into this new high frontier. Where is it going? What can we do there? How can we use it to enhance our lives? How can we use it to benefit us as individuals, businesses, and nations? Can we use this new frontier to solve some of our current problems? The time is ripe; the cycle has run its course and is ready for massive utilization for practical account. What visions do we see unfolding before us?

Easy Access to Space

One of the greatest benefits of developing solar power satellites is the fact that they require the development of a complete space infrastructure, including new low-cost space transportation systems, habitats, and robotic assembly equipment. This infrastructure in turn will make it possible to open space to even more commercial development. Fully reusable heavy-lift space freighters will bring the cost of hauling cargo to orbit down to about one-hundredth to one-thousandth of what it is today using expendable boosters and the Space Shuttle.

During the satellite construction periods, there will be several flights a day. Most of the flights will be dedicated to hauling cargo, but there will also be a need for personnel transportation as well. Passengers could be carried on the freight flights, or there may be separate vehicles just for passengers. In either case transportation costs to space will be within the grasp of nearly any business wanting to haul cargo to space or to an individual desiring to make the trip. Travel to space and back will no longer be restricted to professional astronauts.

In order for the space freighters to load and unload rapidly they must be designed to stow their cargo in containers or on pallets. Each space freighter would be able to carry a number of these containers. They would not need to be as heavy as those made of steel we see on ships, railroads, and trucks, but they would be the same size and have the same type of mounting provisions. Designing the satellite hardware to fit this standard-size shipping container would greatly simplify cargo handling and help reduce cost. It is uncertain at this time how much cargo each space freighter will be designed to carry, but generally speaking the larger the capacity the lower the per-pound cost.

The approach to designing space hardware in the future will be much different from what is done today. Satellites are currently reduced to the minimum size and weight possible because of high transportation costs, and as a result, they themselves become very expensive. With routine, easy, and low-cost access to space, much of that can change. Standard cost-efficient design practices can be used since weight saving will not be as important and size would not be limited. Space freighters could be operated like a cargo or passenger airline operation. If a delivery schedule was missed, there would be another flight the next day with some space available. Like today's air-cargo carriers, a schedule would be maintained even if there wasn't a full load.

Initially there would be only one space line, but as the pace of satellite construction increased it is very likely that other companies would join the industry and establish competition. Competition would bring costs down as technology improved and the industry became mature. If you wanted to take a trip to space, it would not only be possible, but relatively easy. After all, with the early development flights and after the construction of a few satellites, space transports would be well proven for safe routine operation. It should not be unreasonably expensive, probably about the same as a first-class round-the-world ticket on an airline, except that you would be going around the world many times! Passenger compartments would be designed for a comfortable shirt-sleeve environment, with no more need for space suits than for parachutes on passenger airlines today. Flight preparation would be a briefing on emergency procedures and safety precautions such as with today's air travel. There would be no special medical or age restrictions as long as you were in good health. Launch acceleration would be a sensation never before experienced, but laying back in your comfortable, contoured seat, you would probably find the eight-minute trip to the weightlessness of space quite pleasant.

Switchboards in the Sky

Telephones are as common to us as our shoes. We take them for granted. Through the years, they have evolved from black squawky boxes on the kitchen wall to decorator-styled wonders with push-button dialing, automatic dialing, lights, message recorders, and computer control, with even more to come. But with the exception of cellular phones, they are tied to immobile wires leading to switching stations, which then interconnect with other switching stations. These in turn are interconnected with cables, fiberoptics, microwave links, or relay satellites to form the great worldwide spider web of our present-day telephone system. This system is very densely packed in the developed nations but has only spotty coverage in underdeveloped countries, which are unable to afford the massive amounts of capital required for the wire network, switching stations, and telephones.

This fantastic system has two basic drawbacks: it is primarily limited to fixed locations and it is too expensive for more than half the people of the world. The new frontier of space will change that. When large payloads can be transported to space economically and with the ability to assemble large systems in space, a new breed of communications satellites or space platforms can emerge. Most of today's satellites are relay satellites. They receive a signal from the earth and retransmit it to another place on earth. In the case of telephones, the large equipment, antennas, and switching stations are on the ground. If we were to turn these around and put the switching gear, the large sensitive antennas, and powerful transmitters in space on large satellites, that could provide a very different kind of telephone system. The telephone would then become a low-power radio transmitter-receiver requiring no wires. The telephone itself would be the only thing the user would have to buy, and with this small investment, anyone in the world who owned a telephone could use the system simply by paying for the cost of the call.

The first step in this process is already underway. By the late 1990s the first space-based cellular system is scheduled to be in operation. It will soon be followed by others. These low-earth-orbit satellite systems are based on the same concept used in ground-based cellular telephones. Some of them will allow a telephone owner communication access from anyplace on earth. These systems will be launched with our current stable of launch vehicles and are expected to be profitable commercial enterprises. Imagine the potential reductions in cost and increased profits if the satellites could be launched by low-cost space freighters.

Orbiting Vacation Resorts

Up to this point most of what I have talked about is commercial use of space. Another major aspect of space that will open to us will be recreation. The space infrastructure developed to support satellite assembly in space will provide the model for the design of hotels, vacation resorts, restaurants, and hospitals. Can you imagine a more exciting vacation than a week in space, watching the changing panorama of the earth turning beneath you? Spending a couple of hours each day playing that exciting new game called "spaceball?" Relaxing with a good book floating in front of you? Taking a break to look at the stars and other planets through a telescope with no atmospheric interference? Looking down on earth, seeing the Great Wall of China as it wanders along the mountain tops, or looking into the beautiful coral lagoons of the South Pacific, or following the course of the Nile River as it meanders its way from the heart of Africa, or seeing the glint of sunshine reflected from the snow-covered Alps, or seeing the town where you live from a completely different perspective?

Imagine having dinner in a revolving restaurant where the rotation provides a small amount of artificial gravity to keep everything in place on the table and allows the chef's finest concoctions to be served in an elegant manner. As you are seated, it is mid-afternoon on the earth below. When cocktails arrive, you will be passing into the evening shadows watching the glories of a sunset from three hundred miles above the earth's surface. Below you is the darkness of night, and as you look ahead you see first the deep dusk, then the reds and golds and yellows weaving in and out through the great cloud formations. Over to the left is the towering, swirling crown of a large storm with its top still in sunlight. Straight ahead is the brilliant crescent of the earth as the sun dips beneath. Over appetizers the glow of candlelight does not obscure the view of the stars nor a glimpse of the lights of the cities below. After the salad and before the entree, the magnificence of dawn will leave you in awed silence. During dinner you will probably be planning a return trip to this paradise in space. Finally when you reach the end of a perfect dinner, you dawdle over coffee and brandy as another sunset puts your mind at rest. During your leisurely two and a half hour dinner, you have circled the earth one and a half times.

Mining the Cosmos

As we look into the future of a world of ever-expanding population and shrinking natural resources, it is only logical to turn to the new frontier to find new sources. One doesn't think of the emptiness of space as having natural resources, but it is filled with vast quantities of them. Closest to us is the moon and beyond that are the asteroids, the other planets of our solar system, and their moons. Beyond these our imagination fails us—the distances are too great to contemplate. We need not worry, however; our own solar system has an ample supply to satisfy our needs. Our problem is how to gather them for our use.

With space transportation vehicles able to take us above low-earth orbit, we will have access to geosynchronous orbit and beyond. We will be able to return to the moon on a routine basis and utilize the raw materials that it promises. Many dream of building permanent bases and mining the moon. They visualize manufacturing plants to fabricate solar cells and structures to build solar power satellites with materials that do not have to be lifted out of the gravity of earth.

The asteroids are even richer in iron, nickel, and other minerals. They come in many sizes, with new ones being found on a regular basis. They could be mined where they are or they could be moved into orbit around the earth or moon. Bringing larger ones back might be a little dangerous, but the day will likely come when it can be done. I recently listened to a proposal made by a Russian scientist who suggested that one good way to destroy the stock of nuclear missiles left over from the cold war was to use them to move asteroids into earth orbit so they could be used as a source of space-based material. He was very serious about it and had worked out many of the details on how to accomplish the task.

Factories in Space

Putting factories in space is the goal of many visionaries trying to bridge the gap in the development cycle between development and exploitation. They have begged for a Space Station to research the benefits of manufacturing products in zero gravity and the vacuum of space. They have struggled with small experiments on Skylab, Space Shuttle, Russia's space stations, and on unmanned rockets. Several interesting phenomena have been discovered. Two examples are the ability to separate medical drugs in the absence of gravity and the ability to eliminate the convection currents that affect the formation of crystal structures as they cool and solidify.

Progress has been slow, however. The knowledge base is insufficient to paint a clear picture of where the opportunities for space manufacturing will lead. Today it is only feasible to think about space processing for very high-value materials intended for small niche markets. Space Station will be a tremendous help, but it will not be enough to open the door to broad-based manufacturing. That can only happen when access to space is low cost and routine and there is ample power to run the factories. At that time industry can afford to transport large quantities of materials to and from space, and ideas will blossom as industry moves into space.

Manufacturing in space may not reach full maturity until the raw materials it uses actually come from space, thereby avoiding the problems of overcoming gravity. Then space can become the great new industrial center of mankind.

The High Frontier

With the new frontier being the width and breath and magnitude of space itself, it will not only be able to absorb commercial development, but also the migration of people who are looking for a new life, people who are restless, people looking for adventure. Mavericks, explorers, and builders will push on to overcome all obstacles to colonize this new frontier. As our ancestors braved the unknown to follow their dreams, so today there are groups forming to pursue their dreams of totally self-supporting space colonies. These huge colonies would mine raw materials on the moon, grow their own food in a closed ecological cycle, and depend on solar power for energy. They look to space as the escape valve from the spaceship called earth. A place to grow and experience new things, a place to take the next step up the ladder of mankind's development.

Dollars from Space

It was not until 1968 that the concept of solar power satellites providing energy to the earth was proposed. Now as we look about us it is difficult to imagine anything that could provide a more practical benefit to this earth than bringing abundant, low-cost, clean energy from outside our finite world to solve our most fundamental problems. This would be utilization of a frontier in a way that brings enormous good for all mankind. We have already explored the concept in some detail, but let us finally consider the magnitude of the revenue stream that will be pouring from space.

If we assume that electricity will sell for eight cents a kilowatt hour and that the satellites will operate at full capacity as baseload systems, each 1,000 megawatt satellite will generate $675 million a year in revenue. This adds up to $20 billion over a thirty-year period for each satellite. If each satellite lasts one hundred years we will be able to reduce the price of electricity after thirty years when its capital cost has been paid off and the cost of generating electricity drops below two cents a kilowatt hour. If we replaced only half of the current United States generating capacity with solar power satellites, they would generate about $150 billion in revenue per year. If solar power satellites were also used to power half of the current United States transportation market that number would nearly double. If they are used to supply energy to the developing nations of the world the revenue generated by the stream of energy pouring down from the sky could be over *$1 trillion a year*. The magnitude of the numbers is staggering. These satellites will produce enough revenue to pay off the original investment, including the support systems, and return a very handsome profit. After return of the initial investment, the cost of energy from the satellites will drop to the cost of operating and maintaining them. There is no fuel to buy and no more debt to pay. The benefits will come to all of us in the form of very low-cost energy.

As we approach the end of the twentieth century there is no single thing we can do that will have as large an impact on the people of the world during the new century than the development of solar power satellites. They will bring prosperity, an opportunity for the poor nations of the earth to achieve true freedom from want, healing of our environment, and open the vast new frontier of space to all of us.

With the development of solar power satellites we will tap directly into the power of the sun and save the world from impending chaos. There will be hope for the future as we enter the twenty-first century.

The sun. Source of our energy from the beginning of time. Throughout the ages the light of the sun has fueled photosynthesis, freeing oxygen and providing food for the animal kingdom. It supplies the light to grow trees—bringing us wood. Its heat evaporates the oceans to bring the rains that form our rivers and lakes. It causes the winds to blow and brings us warmth and comfort. It took uncounted millions of years for the sun working with the earth to create the coal, oil, and gas we are burning so recklessly today. It is the only fusion reactor in our solar system, yet we turn our backs to it as we huddle in our laboratories fruitlessly trying to duplicate its magnificant power.

The sun. There from the beginning of time, bathing our planet with its life-giving force. We only need to reach out and harness a bit more of that force to create a future for ourselves and our children that is brighter, warmer, cleaner, and more peaceful than any we've yet experienced in our brief time on earth.